THE SHEEP HOUSING HANDBOOK

by

TOM BRYSON

FARMING PRESS LIMITED
WHARFEDALE ROAD, IPSWICH, SUFFOLK

To Jane

First published 1984
ISBN 0 85236 132 7

Made and printed in Great Britain by Bookmag, Henderson Road, Inverness, Scotland.

Contents

Illustrations

PLATES | Page

INTRODUCTION

ALTHOUGH ONLY twenty years ago the idea of housing sheep was treated with some scepticism, sheep housing is not new even in Britain and the other temperate regions of Europe. The Romans housed their sheep flocks as did the ancient Scots during the Romano-British period; they probably shared their accommodation with the livestock. There is evidence that sheep housing at night was extensively practised during the middle ages (Ryder, 1983). The famous chronicle of medieval life in France, *The Book of Hours*, depicts thatched sheep housing with walls extending only halfway to the roof; the builder presumably realised the value of good ventilation. Later in the eighteenth century books on husbandry contained designs for wooden sheep houses. In Scandinavia where winters are prolonged and severe, sheep housing is a long-established tradition; sheep are inwintered for about three months in southern Norway and for up to nine months in the north.

In the United Kingdom until the early 1960s gross margins on the sheep enterprise were low, but so were the fixed costs. There was little incentive to intensify the sheep enterprise and to make such a revolutionary change as sheep housing. The sheep enterprise was not considered worthy of the extra capital investment and there were fears about possible flock health problems. Since the 1960s the pace of change has increased; high land prices, cost inflation and more recently the effect of the Common Agricultural Policy Sheep Meat Regime have stimulated a technical revolution.

New, highly prolific breeds, new vaccines, hormonal control of breeding, artificial insemination and intensive systems to produce six lambs in two years have been accepted by sheep farmers who were once the most resistant to change.

Sheep housing has had an important role in the development of more intensive sheep production in the last twenty years because of the high degree of management control over nutrition and health. In the technical journals, the press, textbooks and advisory publications there is a large body of information on the technical aspects of sheep house design and the management of housed sheep. This book is an attempt to bring together as much of that experience as possible.

There is a tendency in the United Kingdom agricultural industry for farmers to be very concerned with the technical achievement of their farm management. Very often management of financial resources does not reach the high standard of crop and livestock production. Spending decisions involving thousands of pounds are often made on the basis of intuition and experience rather than on a realistic appraisal of the proposed investment. Guidelines for assessing how worthwhile an investment in sheep housing might be are included in the book.

In Britain sheep are kept mainly for meat production; wool is a valuable by-product, but of secondary economic importance. Sheep meat consumption in Britain is low and declining. Younger consumers in the under-35 age group want small, fat-free, waste-free, convenient packs of meat. Traditional joints of lamb do not meet any of these requirements. The general decline in sheep meat consumption could accelerate unless the industry meets the challenge of the market place and supplies what the consumer wants rather than what is convenient to produce.

The challenge of the next twenty years will not be so concerned with technical improvements in production but with more market-orientated production.

Reference

Ryder, M.L. (1983), *Sheep and Man,* Duckworth.

Chapter 1

WHY HOUSE SHEEP?

ECONOMIC PRESSURE has forced the pace of change in the sheep industry. In the lowlands sheep have to compete with intensive cereals and oilseed rape producing gross margins of up to £1,000 a hectare and a high return on capital. Elsewhere the sheep enterprise has to compete with high-yielding dairy cows earning margins over feed and forage in excess of £1,200 a hectare. Where sheep have survived in competition with arable crops on high-grade land, more sheep now produce more lambs on fewer hectares. In only five years from 1977 to 1981 the average stocking rate on MLC recorded farms

Plate 1. With good management it should be possible to produce 1,000 kg live lambs per hectare of lowland grass *Farmers Guardian*

increased by over 20 per cent from 10.4 ewes per hectare to 12.7. The top third of recorded flocks have an overall sheep stocking rate of 16 ewes per hectare. These top third flocks made £3,647 more gross margin per hectare than average flocks on every 20 hectares in 1981. The stocking rate accounted for most of the difference in this case. The output of many lowland flocks could be increased even further by higher stocking rates and better ewe performance. Improved ewe performance can be achieved either by keeping more prolific ewes or by reducing lamb mortality. Both of these goals, higher stocking rates and more lambs sold per ewe, can be achieved with the help of winter housing.

The case for housing is less clear cut in the hills and uplands. Competition from more profitable enterprises is unlikely and in many cases the levels of output even after housing would be too low to service the capital investment. However, in a recent survey of hill sheep housing in the north of England farmers claimed better productivity, ewes were fitter, lambs were stronger and poaching of valuable inbye land was reduced. The economic advantages of housing hill sheep have not been clearly established but inwintering is becoming more widely practised.

More ewes on fewer hectares

It is technically feasible and profitable with good management for one hectare of lowland grass to produce 1,000 kg of live lamb each year. Most sheep systems in the lowlands produce less than 500 kg a year. A tonne of ewe liveweight per hectare is needed to grow a tonne of live lamb. An overall stocking rate of fifteen Mule or Greyface ewes per hectare mated to a Suffolk ram and rearing 1.6 lambs of 40 kg liveweight a ewe could reach the 1,000 kg a hectare target. Inwintering, by reducing winter poaching and increasing lamb survival, is an important factor in reaching an output of this level. Not only does this affect the profitability of the sheep flock directly, but higher stocking rates release more land for alternative enterprises. More wheat, oilseed rape or early bite for the dairy herd have an effect on the whole farm profit.

Silage production is much easier to integrate into an inten-

Table 1.1. Lambs needed to produce 1,000 kg lamb liveweight per hectare

Av. lamb liveweight (kg)	*Lambs per hectare*
30	33.3
35	28.6
40	25.0
45	22.2

Source: ADAS, 1981.

sive grazing system than haymaking. Although it is possible to feed silage to outwintered ewes in big bales or from self-feed clamps, wastage is high and poaching is a problem. Housing does make the change from hay to silage easier as wastage is low and feeding can be mechanised. On the arable farm untreated straw can be fed to housed sheep as a substitute for hay, again pushing up the stocking rate and releasing arable hectares.

More grass and earlier grass

The areas of the country with the greatest grassland production potential are usually most prone to poaching in winter. Although there is little grass growth in most winters, removal of sheep from the fields has a significant effect on the earliness of growth and dry matter yield in April and May. Most damage is done by grazing between January and the end of March. Several attempts have been made to measure the effect of late winter grazing (Table 1.2). Reductions in productivity range from 8 per cent to as high as 50 per cent in some years.

Table 1.2. The effect of winter grazing on grass production in spring

Source	*Percentage reduction in dry matter yield*
Frame 1975	−8 to −16%
McG. Cooper 1966	−20%
Lockhart 1969	−50%
Bastiman & Kneale 1974	−16%

Late winter grazing from January to the end of March appears to cause the biggest reduction in grass yield. The effect is greatest from early April to mid-May, often the hungriest time of year on a livestock farm. Ungrazed swards provide earlier grazing and more grass. As the season progresses the effect of winter grazing on grass yield is reduced. In most years the ungrazed fields respond to fertiliser more quickly and produce about 10–15 per cent more dry matter during the April–May period.

Winter grazing of silage fields can also have repercussions on first-cut silage yields and early bite for dairy cows on the same farm. After housing 1,000 ewes in January 1980 I expected to make about 15 per cent more first-cut silage. The effect of inwintering was much more dramatic than I had expected. Yields of first-cut silage dry matter were 26 per cent higher than in the previous year. Fertiliser rates and cutting dates were the same in both years, the main difference in 1980 was that the ewes had been housed from January until lambing.

Table 1.3. Silage production Lazonby Estate farms 1979–80

	1979	*1980*
Silage area (ha)	82	99
1st cut silage yield (tonnes)	1,300	2,000
DM yield (tonnes/ha)	3.98	5.05

More lambs per ewe

Apart from helping to increase stocking rates, housing should lead to increased ewe output. A higher degree of management control is possible when ewes are housed. Keeping more prolific breeds such as the Cambridge and Finn crosses, perhaps producing three crops of lambs in two years, becomes feasible with housing. Even the more traditional breeds and crosses — Mules, Greyfaces and Scottish halfbreds — have a greater production potential than is usually realised in practice. The nutrition of housed sheep can be matched more accurately to ewe condition and expected lambing date. More accurate feeding means stronger lambs, more milk, better lamb survival, higher weaning weights and more lambs sold fat off grass.

The effect of reduced lamb mortality will vary between farms depending on the severity of the climate and the degree of shepherding expertise. Lambing percentage is largely determined in the period during and immediately after mating; it depends on the number of eggs shed and early embryo survival. But commercial and experimental experience suggests that an increase of 10 per cent in the number of lambs weaned can be realistically expected when ewes are housed. Under severe weather conditions losses well in excess of 10 per cent could be prevented.

Much of the early development work on inwintering was done at Great House EHF in Lancashire. Inwintered ewes had a consistently higher lambing percentage than outwintered ewes over a four-year period (Table 1.4). Each year the inwintered and outwintered sheep were selected at random to avoid bias and the inwintered ewes averaged 16 per cent more lambs.

Table 1.4. The effect of inwintering on lambing percentage

	1963–4	*1964–5*	*1965–6*	*1966–7*	*Average*
Inwintered	152	145	180	168	162
Outwintered	139	134	163	146	146

Source: Bastiman & Kneale, 1972.

Estimates of lamb mortality vary. Lowland flocks recorded by the MLC in 1974 and 1975 had a mortality rate between 12 and 14 per cent. Other estimates range as high as 15 to 20 per cent, and it has been estimated that three million lambs die within three days of birth every year. At 1983 prices the slaughter value of these lambs could be worth £100 million on a national scale. Reducing lamb mortality from 15 or 20 per cent to 5 or 10 per cent goes a long way to paying for a sheep house.

After the blizzards and deep snow of the 1979–80 winter, I estimated that we lost over 20 per cent of lambs born alive because of the weather. Housing would certainly have halved

Plate 2. Feeding sheep outdoors in winter *Farmers Guardian*

Plate 3. Housing improves the shepherd's working conditions in winter

these losses. In the following winter when the ewes were housed from January until the end of March, the number of lambs weaned was 13 per cent higher than in the previous year. The lambing percentage increased from 143 per cent in 1979 to 156 per cent in 1980. Ewe mortality fell by one per cent from 4.5 to 3.5 per cent. Although this was an improvement on the previous year's results, it should have been even higher. Poor silage and poor body condition of the ewes led to a lack of colostrum and losses of small weak lambs. Housing alone does not guarantee success. Management has to be even better when the flock is housed.

Better working conditions

Sheep housing is much more convenient than outwintering. Anyone who has fed outwintered ewes in rain or snow after travelling through muddy gateways and rutted fields will appreciate the more pleasant working conditions. The sheep are concentrated near at hand when their need for accurate feeding and close attention is greatest. Siting is a matter of choice; proximity to the farm house, a hard road, feed stores or other buildings to be used in conjunction with the sheep house can all add convenience.

Both sheep and shepherd are protected from the weather at lambing. Close supervision can be given twenty-four hours a day at the peak of lambing. Electric lights and covered pens do not necessarily make the shepherd's job easier but they do enable him to do it better. Inwintered sheep are more easily caught with less stress for prompt attention. If the handling pens are incorporated in the layout or are adjacent to the house, routine dosing, vaccination and foot treatment are less time consuming. Mechanised silage feeding becomes a possibility with housing. This may mean that less labour is committed to the sheep enterprise or that more sheep can be handled by existing labour. If experienced casual labour is available at lambing, one man should be capable of managing 800 to 1,000 ewes.

Protection from predators

Sheep worrying by dogs is not only confined to farms on the urban fringe but it is also widespread in the countryside. In

Plate 4. Mechanised silage feeding is possible when sheep are housed
Farmers Guardian

the hills and uplands foxes and carrion crows occasionally wreak havoc among new-born lambs outside. Housed ewes are protected from these predators in the vulnerable pre-lambing period and the lambs are protected immediately after birth.

Lower replacement costs

Flock replacement costs increased by 80 per cent in the lowlands and 152 per cent in the uplands between 1977 and 1981. Even after allowing for 12 per cent annual inflation there has been a real increase of 30 per cent in the lowlands and just over 100 per cent in the uplands (Table 1.5).

Table 1.5. Increase in flock replacement cost — £s per ewe 1977–81

	1977	*1978*	*1979*	*1980*	*1981*
Lowland flocks	4.0	5.7	6.1	6.5	7.2
Upland flocks	2.5	3.5	4.9	4.6	6.3

Source: MLC, 1982.

If an extra crop of lambs is obtained from a ewe that would otherwise be drafted, replacement costs can be cut substantially. In practice older ewes have shown remarkably improved performance after housing.

At the other end of the flock age range, breeding from ewe lambs can also reduce flock replacement costs. There is an added bonus from breeding ewe lambs as they tend to be more prolific in subsequent years. However, ewe lambs are still growing and adequate nutrition is essential. They also need very close supervision at lambing because of their excitability and lack of maternal experience. Breeding from ewe lambs requires a high degree of skill and attention to detail; housing makes the job easier.

Alternative uses for sheep housing

Most sheep houses are only occupied for three to four months of the year. If a profitable use can be found for the building during the remaining eight months, the annual charge per ewe will be less. Using the house as a feedlot to finish store lambs is one possibility but this does depend on the relative cost of forages. In some situations silage or arable by-products for housed lambs may be more profitable than grazed forage crops and can increase stocking rates even further. Temporary storage of hay, straw or machinery are other possibilities. At least one enterprising farmer has used a polythene tunnel sheep house to produce early strawberries in peat gro-bags on pallets.

Sheep housing is expensive

Not many existing farm buildings are suitable for sheep without alteration to improve ventilation or to give more protection from the weather. Where it is feasible, conversions are usually the cheapest option. On the other hand, package-deal portal-frame buildings can cost up to ten times more per ewe housed. The building is not the only expense: site preparation, water supply, electricity supply, hard roads, feeding equipment and extra feed storage can all add to the cost. Unless management is good, poorer ewe performance could result from housing and the capital invested would not be recouped.

There are some intensive outwintering systems which could be more economic than housing under some conditions. The East of Scotland College of Agriculture developed intensive winter stocking of ewes, up to thirty-five per hectare from late January until lambing in mid-March. This achieves many of the benefits of housing without the capital expenditure. There is more grass at lambing, labour for feeding and shepherding is reduced and temporary shelter can be provided for lambing. On the other hand, well-drained fields are needed and there is no protection from weather; the wintering area has to be conserved or ploughed up in spring if it is badly poached.

As the capital investment is high, levels of output must be correspondingly high to service the investment over a relatively short period. However, return on capital may not necessarily be the most important factor on which a farmer judges an investment. The amount of extra income generated may be more important. Generally it is more prudent to look for a return on marginal capital investment at least equal to that from alternative investments.

Variable costs are higher with housed sheep; straw for bedding and handling of wastes add to the cost. The ewes eat more. Extra feed needed depends on the length of the housed period and how much they would have been fed if outwintered. The intensive outwintering system probably requires more hay because wastage is greater. Concentrate usage should be the same as when the ewes are outwintered. Housed ewes eat 10–15 per cent more hay and silage and slightly more if shorn at housing.

Disease could be a problem

If animals are closely confined in large numbers for relatively long periods, the risk of disease is greater. But good design for adequate ventilation cuts respiratory problems and a dry, well-drained floor keeps foot problems to a minimum. A comprehensive, year-round flock health programme is suggested in Chapter 8, which should be modified according to the farm and after taking veterinary advice. Problems do occur even with the best of planning and preventative medicine. The risk of disease is greatest around lambing. The build-up of pathogenic bacteria in a sheep house and lambing pens can be very rapid. An investigation in the east of Scotland showed that the concentration of *Escherichia coli* during lambing increased 48 times in outdoor pens and 177 times in indoor pens.

Sheep housing checklist

The advantages of housing:

- A higher degree of management control is possible.
- Higher stocking rates are easier to achieve.
- Grass productivity is higher in spring.
- Ewe performance is improved.
- Lamb mortality is reduced.
- Working conditions are better.
- Labour can be used more effectively.
- Feeding can be mechanised.
- Sheep are protected from predators and bad weather conditions.
- Flock replacement costs can be reduced.

The disadvantages of housing:

- Capital cost is high.
- Extra variable costs are incurred.
- Better all-round management is needed.
- The risk of disease could be higher if attention is not paid to detail.
- The value of the fleece may be depreciated.

References

ADAS (1981), 'A system for lowland sheep', *Booklet 2322*.

Bastiman, B. and Williams, D.B. (1973), 'Inwintering of ewes. Part I. The effect of housing', *Expl Husb. 24*, 1–6.

Bastiman, B. and Kneale, W.A. (1974), 'Inwintering of ewes. Part III. The effect on the sward', *Expl Husb. 27*.

Frame, J. (1976), 'The effect of winter grazing by sheep on spring and early summer pasture production', *J. Br. Grassld Soc. 25*, 167–71.

Lockhart, D.A.S., Herriot, J.B.D., Cunningham, J.M.M. and Heddle, P.G. (1969), 'The effects of winter grazing on subsequent production from pasture', *J. Br. Grassld Soc. 24*, 146–50.

Maumid, B.A., Duffell, S.J. and Winkler, C.E. (1980), 'Lamb mortality in relation to prolificacy', *Expl Husb. 36*, 99–112.

McG. Cooper, M. (1966), *Outlook on Agriculture 5* (2).

MLC (1982), *Commercial Sheep Production Yearbook*, 1981–2.

Chapter 2

HOUSING HILL SHEEP

APPROXIMATELY ONE-THIRD of the agricultural land surface of Great Britain is classed as rough grazing, with two-thirds of this in Scotland. Where rough grazing occurs at high altitude, productivity of the land is low because of considerable climatic and physical limitations. Poor thin soils, difficult terrain, severe climate and remoteness of hill sheep farms combine to generate only low output and low profits, which discourage capital investment. In the lowlands where land prices are high and there are high output alternatives to sheep farming the economic advantages of inwintering are obvious; but in the

Plate 5. Hill sheep farming is characterised by low output and low profits. Housing may have a place where productivity has been increased after land improvement *Farmers Guardian*

hills return on investment is generally low and highly variable. Housing may not have a place on many hill sheep farms, but where land has been reclaimed and productivity increased to a satisfactory level housing could lead to further increases in output and profitability.

Forty per cent of the agricultural land in Wales is over 250 metres above sea level (Jones, 1967). In Scotland there are over 1 million hectares above 450 metres. Hill sheep farming is the most important economic activity in these areas of severe climate. For every 100 metres above sea level the temperature drops by 0.6°C, wind speed increases with altitude as do rainfall, snow and cloud cover. Conditions for plant growth therefore deteriorate very rapidly with increasing altitude and distance from the modifying effects of the sea.

Table 2.1. Dry matter yields of hill pasture types

Vegetation	*Production kg DM/ha year*
Agrostis + Agrostis/fescue	2,240–2,800
Molinia/Nardus	1,120–2,240
Heather	1,120–2,800
Lowland perennial ryegrass (330-560kg N/ha)	11,200–16,800

Source: Eadie, 1970.

The nutritive quality of unimproved hill pasture also tends to be low, leading to reduced nutrient intake by the grazing animal. Heather, rushes, sedges and coarse hill grasses have a lower digestibility than sown species and do not regrow rapidly after grazing. Where the herbage is ungrazed there is a steady decline in quality with maturity, when stocking rates and grazing pressure are low.

Hill pastures tend to be undergrazed during the summer because stock numbers are limited by the winter carrying capacity of the farm. Almost 75 per cent of hill ewes are stocked at between 1 and 2 hectares a ewe (Table 2.2).

Low sheep stocking rates are not the only problem. Individual ewe performance is poor, weaning percentages can be as low as 60 with an average of about 100 (MLC). These levels are a long way below the genetic potential of the two most

Table 2.2. Stocking rates of hill sheep in Scotland

Rough grazing per ewe and gimmer (ha)	*Total hill sheep %*	
0–0.4	0.6	
0.4–0.81	6.2	
0.81–1.21	34.7	} 77.1 (0.81–2.43)
1.21–1.62	20.6	
1.62–2.02	11.6	
2.02–2.43	10.2	
2.43–2.83	5.8	
2.83–3.24	1.8	
3.24–3.64	3.6	
3.64–4.05	1.1	
74.05	3.8	

Source: Eadie, 1970.

numerous hill breeds, the Scottish Blackface and the Welsh Mountain, which between them account for over 60 per cent of the 7 million hill ewes in Great Britain. Better nutrition is the key to realising more of the potential of hill ewes. Improved overall nutrition with better body condition in the ewe results in more lambs weaned at greater weights. Weaning rates are low with a range from 18 kg liveweight to over 27 kg but despite the limitations of breed size and environment there is more potential for improvement. Low numbers of lightweight lambs at weaning means low output of saleable meat for each hectare of hill land, again there is variation in performance from less than 4 kg lamb liveweight a hectare to more than 34 kg a hectare. Compare this with the 1,000 kg lamb liveweight a hectare achieved on intensively stocked lowland farms. A high proportion of hill lambs are sold as stores. Until recently with the introduction of the EEC sheep regime the store market was very unpredictable from year to year, but hill farmers were under pressure to sell because of winter coming and a shortage of forage.

Low output is the major economic problem of hill sheep farming; low output results in low profits leaving only a small or non-existent surplus for investment which could be used to increase output. However, within Great Britain there are wide differences in hill sheep farming systems, differences that depend on soil, herbage and location. Perhaps the most important factor is the ratio of hill grazing to inbye land (Univ. Newcastle-upon-Tyne A.A.U., 1970). Where the proportion of inbye to hill grazing is high (1:15 or less) there is more winter keep, a higher stocking rate is possible as is a better level of nutrition, more lambs are weaned, ewe mortality is low, replacement rate is low and more lambs are sold fat. At progressively lower rates of inbye to hill grazing (1:20 or more) output is correspondingly reduced.

Hill farms with ewe housing also have an advantage. Where housing is available stocking rates are less limited by the winter carrying capacity of the farm, and nutrition and lambing percentages can be improved. Fewer ewes and lambs die and an extra crop of lambs may even be obtained from sheep that would otherwise be drafted. In some cases hogg wintering costs may be lower when the animals are housed. Cattle numbers might also be increased when ewes are inwintered as silage or hay crops are heavier (Elsy, 1980).

Hill sheep production systems can be put into three rather broad categories: those with common grazings; those with enclosed hill and limited inbye, as in many parts of Scotland; and enclosed hill with a substantial proportion of inbye, as in Wales.

Common hill grazings

Hill farms in this category are probably the most difficult to improve as a high degree of co-operation and agreement are needed before land can be reclaimed. The lack of fences and grazing management are further obstacles to sustained improvement by reseeding. Management at tupping and lambing is difficult thus, ewe mortality can be as high as 10 per cent per annum and lambing rates as low as 50 per cent have been recorded. There are therefore fewer lambs and draft ewes to sell. All sales are of store lambs and output in these areas is low and unstable. Investment in sheep housing should have a lower priority than reseeding and fencing in these cases.

Enclosed hill and some inbye (greater than 20:1)

There are opportunities for land improvement, fencing and closer management with this system. Ewe mortality is usually about 5 per cent and lambing percentages range from 85 to

100. Because there is limited inbye and usually competition from cattle for forage there are no opportunities to fatten lambs, therefore draft ewes and lambs are sold in store condition. Quite small amounts of capital invested in land improvement are likely to generate considerable increases in output and cash income quickly and with little reliance on long-term borrowing. Large capital investments, as for example in housing, can cause cash-flow problems. Increased animal performance through better pasture and/or more sheep should have priority for capital in this type of system. The strategy should be to improve net farm income with relatively low levels of investment. Once this investment starts to generate greater profits then more substantial investments such as housing can be considered.

Enclosed hill and high proportion of inbye (15:1)

This type of hill sheep farm is closer to the 'upland' situation and has fewer limitations than the previous two systems. Winter carrying capacity is high, which in turn increases the summer stocking rate and efficiency of utilisation of hill pasture. A high plane of nutrition for the ewes all year round results in more lambs at weaning and lambs that grow faster to sell at higher prices. There may also be opportunities to fatten a proportion of the lambs on forage crops, as in the Redesdale system of hill land improvement. A proportion of the ewes (25-50 per cent) may even be used to produce Mule, Greyface or Halfbred lambs. Farms in this category generally have a much higher output than the previous two types and capital may well have been invested in reseeding, fencing, cattle housing and sheep housing. Where sheep housing has been erected it can be used for a large part of the year, from mid-January until mid-May for in-lamb ewes, for shearing from mid-June until mid-July and then for intensive lamb fattening from late September until January.

Three broad categories of hill sheep farming have been identified, and it must be emphasised that these are very broad groups. This wide range of physical difficulties and technical problems in the hills is also reflected in the financial results from MLC recorded flocks. In the five years up to 1982 lamb sales per ewe and draft ewe prices increased, but when

Table 2.3. Results for 51 hill flocks, 1982

Financial results	*Average (£ per ewe)*	*Top third (£ per ewe)*	*% Difference*
Output			
Lamb sales	19.54	23.66	+21
Sale of draft ewes	5.56	7.01	+26
Wool sales	1.42	1.53	+8
Hill ewe subsidy and premium	7.62	7.41	−3
Gross returns	34.14	39.61	+16
less flock replacements (rams only)	1.58	1.36	−1
Output	32.56	38.25	+17
Variable costs			
Concentrates	2.44	1.61	−34
Purchased forage	0.43	0.24	−44
Forage variable costs	1.44	1.34	−7
Total feed & forage	4.31	3.19	−26
Vet and medicine	1.52	1.30	−1
Miscellaneous & transport	0.51	0.54	+6
Total variable costs	6.34	5.03	−21
GROSS MARGIN (output-variable costs)	26.22	33.22	+27

Source: MLC, 1983.

the figures are adjusted for inflation there is no difference between 1981 and 1982. During the period extra output came from higher lamb and draft ewe prices rather than more lambs. Lambs reared per ewe was relatively static at 1.0 over the five-year period. However, when the 'top third' flocks are compared with 'average flocks' there are important differences.

Gross returns from the 'top third' flocks were 16 per cent higher than average because of higher lamb and draft ewe prices (Table 2.3). The 'top third' producers fed less concentrates and less purchased forage in general and their variable costs were 20 per cent less than average; giving a 27 per cent higher gross margin. The better farms sold 13 per cent more lambs, with a higher proportion sold fat (Table 2.4).

There is therefore some scope to improve individual ewe performance in the hills, while simultaneously increasing ewe

Table 2.4. Physical results from hill flocks, 1982

	Average	*Top third*	*%Difference*
Lambs reared (per 100 ewes to ram)	100	113	+13
Lambs sold finished (percentage)	21	26	+5

Source: MLC, 1983.

numbers and farm output. Hill pasture improvement is the key to improved output and economic performance. Better individual ewe performance is made up of the following factors:

- Higher percentage of lambs reared.
- Lamb sale price — lamb fattening.
- Draft ewe price.
- Value of wool clip.
- Ewe mortality.
- Longevity.

The number of lambs reared per 100 ewes put to the ram and the sale price of the lambs has the greatest effect on output (Tables 2.4 and 2.5).

Table 2.5. Effect on output of increasing lamb rearing percentage with different lamb prices

%Lambs reared	*Output per ewe Average sale price per lamb(£)* 25	27	29
80	20	21.60	23.20
100	25	27.00	29.00
120	30	32.40	34.80

Note: See Appendix 1 for calculation of output per ewe.

Both factors — lambing percentage and lamb sale price — are affected to a high degree by nutrition. As has been seen, unimproved semi-natural vegetation tends to be undergrazed, low in digestibility and low in calcium, magnesium and phosphorous content. The improvement of these pastures with lime and reseeding increases digestibility and utilisation, but grazing must be controlled. Nutrition prior to mating and during lactation can be improved resulting in more lambs reared and sold at higher prices. After land improvement there is a greater need to improve nutrition in late pregnancy — inwintering allows a greater degree of control over nutrition. Housed ewes with more milk, giving birth to more lambs, and under closer supervision at lambing could easily rear more lambs (Evans, 1980).

Table 2.6. Performance of 1,300-ewe flock after housing
(Welsh mountain and hardy speckled ewes; regular ages; maximum lambing percentage before housing 107).

Year	*Ewes to tup*	*Lambing %*	*Lambs sold or retained*	*Extra lambs*	*Value of extra lambs @ £20 ea.*
1978	1,302	135	1,751	364	7,280
1979*	1,301	123	1,602	208	4,160
1980	1,295	134	1,736	349	6,980

* Hard winter
Source: Evans, 1980.

Year-round grazing systems supplemented with land improvement should be the first priority. Very large increases in output are needed to pay for the large capital investment in sheep housing and associated increases in variable costs.

Plate 6. Inwintered hill ewes in Corwen, N. Wales *Farmers Guardian*

Investment in housing should only be considered after pasture improvement has boosted net farm income and cash flow; only in unusual circumstances would housing have top priority.

The benefits of inwintering hill ewes in northern England were described by Elsy in 1979, however no economic evaluation of inwintering was attempted. Thirty-one hill sheep farms were visited in order to identify the benefits of housing. Farmers interviewed made the following observations:

- Inwintering led to high productivity due to better ewe body condition, heavier lambs at birth and more lambs reared.
- Labour requirement was significantly reduced and a higher standard of management made possible.
- Poaching of valuable inbye was much reduced in late winter.

Sheep housing at Redesdale

Housing for 320 Blackface ewes was erected at Redesdale EHF, Otterburn, Northumberland in 1970. Since then it has

Plate 7. Inwintered ewe lambs at Redesdale EHF, Northumberland

been possible to monitor the costs and returns to this investment. Despite a relatively low capital cost of £4,650 in 1970 and a cost per ewe of £14.53p over a twenty-year repayment period with the interest rate at 18 per cent, the annual charge is £870 or £2.72 a ewe. This extra fixed cost must be recouped by higher output. Variable costs are also higher after housing. The ewes are housed at the end of February 6–7 weeks before the start of lambing. During the eleven-year period from 1970 to 1981 the average feed consumption was 77 kg of hay and 37 kg of concentrates. At 1981 prices this was worth £7.40. There is also an outwintered Blackface flock at Redesdale stocked at a similar rate (2.2 ewe/ha). Average feed consumption of the outwintered flock is 25 kg of hay and 18 kg of concentrates costing £3.40. One and a half bales of straw are needed to bed each of the housed ewes. The housing cost was £7.22 a ewe more than outwintering (Table 2.7).

Table 2.7. Cost of housing per ewe at Redesdale EHF

	Housed flock		*Outwintered flock*	
	Kg	*Value £*	*Kg*	*Value £*
Hay	77	7.40	25	3.40
Concentrates	37		18	
Straw	25	0.50	—	—
Building	—	2.72	—	—
		10.62		3.40

Source: ADAS, 1983.

Lower lamb mortality does go some way towards paying for the extra cost of housing as extra lambs reared can range from 4 per cent (Redesdale, 1982) to as high as 28 per cent (Evans, 1980). An average value of 8–10 per cent extra lambs after housing is probably realistic (Armstrong, 1984) and this could be much higher in severe winters such as the winter of 1978-79. Ewe mortality could also be reduced by inwintering.

At Redesdale the extra cost of housing could be paid for if the ewes produced 10 per cent more lambs for every 100 ewes mated, and the number of ewes could also be increased by 10 per cent. In fact, since the ewes were housed at Redesdale numbers have risen from 271 to 361; 34 per cent more. As there are forty more than the building can hold these extra sheep are bedded on straw outside. On average 4 per cent more lambs have been reared since housing. (This relatively low figure probably reflects the high standard of management of the outwintered flock.) More ewes and extra lambs worth £2,912 at £26 each have more than paid for the extra cost of housing (Table 2.8).

Table 2.8. Financial effects of housing hill ewes — Redesdale EHF

	Lambs sold	*Average price*	*Value £*
After housing	405	26	10,530
Before housing	293	26	7,618
Difference	112	—	2,912

Source: ADAS, 1983.

The extra income generated by more ewes and lambs exceeded the extra cost of housing in 1981 by approximately £300, but wool clip (£129) and extra hill sheep subsidy (£693) from 91 ewes should be added to the output.

Additional benefit has been obtained from the Redesdale sheep house by:

- Fattening lambs indoors.
- Obtaining an extra lamb crop from draft ewes.
- Use as a shearing shed.
- Hay/straw storage.
- Reduced winter poaching/grazing of improved pasture.

At the Hill Farming Research Organisation (HFRO) farm at Sourhope in the Scottish borders the sheep house, which was built in 1971, is used profitably for nine months of the year:

Mid-January to mid-May — Ewes housed
Mid-June to mid-July — Shearing (ewes housed at night)
Late Sept to January — Indoor lamb fattening.

Fattening lambs indoors

At the Sourhope research station of HFRO about 900–950 lambs are fattened indoors each winter. If these smaller lambs were sold as stores it would be at a disadvantage as indoor

Plate 8. Scottish Blackface wether lambs fattened on silage, barley and fishmeal *Farmers Guardian*

fattening leaves an extra margin of £4–5 a lamb. Add the extra margin from lamb fattening to the economic advantages of housing the ewe flock and the combined extra earnings should more than pay for the capital cost of the house.

Taking a small hill lamb from store to fat condition on a concentrate diet while kept indoors can be profitable. Feeding lambs from September until January is done against a background of rising market price. At Redesdale EHF small lambs and incompletely castrated 'chasers' are housed towards the end of October.

As with all dietary changes, the move from forage-based to concentrated diet should be gradual and must start before the lambs are housed. Start with a small quantity of concentrate and plenty of troughs. It may be advisable to hold the group of lambs near the troughs with a dog until they get used to the idea of eating their ration. Sugar beet pulp, dried grass pellets and flaked maize are all highly palatable ration ingredients which encourage feeding. After about two weeks lambs should be eating 220 g a day, the ration is then increased in steps:

Weeks before housing	Concentrates g/day
4	50–100
3	100–220
2	220–340
1*	340–500

* Two feeds a day may be safer than one.

The concentrate diet can be compounded from a variety of ingredients, whole grain is preferable to rolled grain and suitable rations are described in Chapter 8.

If the building which is to be used for the lambs has adequate ventilation for adult sheep then the lambs will not suffer. Allow 0.35 m^2 for a 30 kg lamb, and up to 0.5 m^2 for heavier lambs. Not less than 250–300 mm of trough space is needed of each lamb.

Targets for indoor lamb fattening

Weight at start of feeding period	30 kg
Weight when fat	34.5 kg
Dressed carcass weight at sale	15.2 kg
Growth rate	1.0 kg a week
Feed consumption	0.9 kg a day

	£
Sale of lambs 15.2 kg DCW @ £1.90/kg	28.88
Less value of small store lambs	20.06
NET OUTPUT	8.82
Variable costs	£
Concentrates 27 kg @ 13p/kg	3.51
Straw bedding	0.20
Vet. and Med.	0.30
	4.01
GROSS MARGIN	4.81

Using these target figures the likely costs and returns of the system can be accurately estimated in advance. An extra margin of about £5 for keeping the lamb a further 4–6 weeks must be attractive.

Lambs from draft ewes

Every year large numbers of hill ewes are drafted from hill farms after producing three or four lamb crops. Severe weather and scarce forage of low quality limit the performance of these older sheep, but their potential can be realised on farms with better grazing and less rigorous climate. In the north of England and Scotland draft hill ewes are a good investment for the lowland farmer who produces cross-bred lambs using Blue Faced Leicester, Border Leicester or Teeswater rams. Housing gives the hill farmer the opportunity to keep these ewes for an extra year when their potential prolificacy is still high. There is also the possibility of selling cross-bred lambs where summer nutrition is adequate for good growth rates.

Results from MLC recorded flocks show that draft hill ewes can produce heavy lamb crops. Only the strongest ewes should be selected and they must be in good condition at tupping (condition score 2.5–3). A draft Swaledale ewe should weigh 47–50 kg at tupping with a condition score of 2.5–3.

Table 2.9. Performance of draft Swaledale ewe flocks in different condition at tupping

	Good condition	*Moderate condition*
Ave. ewe condition score	2.8	2.0
Ave. weight at tupping (kg)	48	44.5
Barren ewes (per 100 ewes mated)	4	8
Live lambs born (per 100 ewes mated)	141	119
Lambs reared (per 100 ewes mated)	130	106

Source: MLC, 1976.

Inwintering hoggs (ewe lambs)

Severe weather and the low winter carrying capacity of many hill farms makes it very difficult or even impossible to over-winter ewe lambs outdoors on the hills. It has been customary to transfer ewe lambs to lowland farms for the winter (away wintering), usually from mid-November until the end of April.

During the last decade, lowland farmers have tended to become more specialised either in arable or dairy enterprises. On arable farms fences and hedges which have not been removed have fallen into disrepair and dairy farmers do not want winter grazing by sheep to prejudice their chances of an early bite. It has become progressively more difficult for the hill farmer to find wintering for his lambs. At the same time both wintering and transport have become more expensive. Wintering ewe lambs at home, indoors on slats, has had a higher priority than ewe housing. Away wintering from October 1983 until April 1984 cost in the region of £10 a lamb including transport.

Hay and purchased concentrates needed to add 15-16 per cent body weight to a 30 kg lamb over the winter would cost about £7.50. After feed costs this only leaves £2.50 a head to pay for the extra cost of housing. However, as ewe lambs only need 0.6–0.75 m^2 each this does help to reduce the cost. There is also the added advantage over away wintering in that the management of the lambs is under the farmers' control. Until the ewe lambs were housed at Redesdale they were hand fed and outwintered. After housing, feeding only took ten to fifteen minutes a day compared with one and a half hours when outwintered.

Housing twin-bearing ewes

It is now possible to determine with a high degree of accuracy (98 per cent) which ewes are carrying twins. This is achieved by ultrasonic scanning during the third month of pregnancy. Once ewes carrying twins can be identified the economics of housing hill sheep change dramatically. Mortality rates of twin lambs born on the open hill are high, estimates vary from 10 to 20 per cent or even higher in severe weather. The vast majority of these lambs could be saved if they were born indoors. Ewes could also be fed more accurately in late pregnancy to ensure greater lamb birth weights and more milk. Hill ewes with access to more pasture at tupping and in good condition should be capable of giving birth to 140–150 lambs for every 100 ewes mated. A lambing rate of 150 per cent means that half the ewes have singles and half have twins. That half of the flock carrying twins could be housed in January and fed accordingly. Ewes carrying singles, on the other hand, could stay on the hill through to lambing with minimal supplementation from feed blocks and hay during periods of snow cover.

At lambing the shepherding effort could be concentrated on the housed ewes to great effect. Hill shepherds spend a large amount of their time around lambing moving ewes with twins from the hill to more productive sheltered pasture. As anyone who has ever done it knows this is a frustrating and time-consuming occupation. By housing only twin-bearing ewes all the advantages of inwintering can be bought at half the cost.

Housing hill sheep checklist

- Invest in pasture improvement and more ewes to increase income before considering housing.
- Hoggs (ewe lambs) may have the highest priority housing where winter keep is difficult and expensive.
- Plan to make the maximum use of any investment in housing by fattening lambs indoors, housing draft ewes, storing hay and using it as a shearing shed.
- Consider housing twin-bearing ewes only.
- If an open-fronted (Rosemaund-type) house is used siting is important. Wind, rain and exposure at the forage troughs can seriously reduce feed intake.
- Ensure that any building is designed to survive severe gales and snowloading; take professional advice if necessary.

References

ADAS (1982), *Redesdale EHF Annual Review.*

Armstrong, R.D. (1984), (Personal communication.)

Eadie, J. (1971), 'Efficiency of hill sheep production systems', in *Potential Crop Production*, Eds. Wareing, P.F. & Cooper, J.P., Heinemann, 239-49.

Elsy, D.K. (1979), 'Housing of hill sheep in northern region', ADAS Rpt.

University of Newcastle-upon-Tyne — Agricultural Adjustment Unit (1970), 'Hill sheep farming today and tomorrow', *Workshop Rpt. No. 13.*

Evans, A. (1980), 'A philosophy for land improvement and agricultural production in the hills and uplands', *Proceedings of Occasional Symposium No. 12*, The British Grassland Society, Edinburgh, 35-40.

JONES, L.L.I. (1967), 'Studies on hill land in Wales', *Welsh Plant Breeding Station Tech. Bull. 2*.

MLC (1976), *More Lambs From Draft Ewes*.

MLC (1983), *Sheep Production Yearbook 1982*.

Chapter 3

WILL HOUSING BE WORTHWHILE?

MANY FARMERS make their investment decisions intuitively by a combination of business acumen and experience. However, building costs are high (Figure 3.1) and so is the cost of borrowing. A wrong decision could be disastrous for some farm businesses. The proposals may 'feel' right and you may want to go ahead but check your hunch with an objective test. In any case, if you are borrowing money the lender will want to know whether the investment appears worthwhile and if there will be enough cash to service the loan. Remember too,

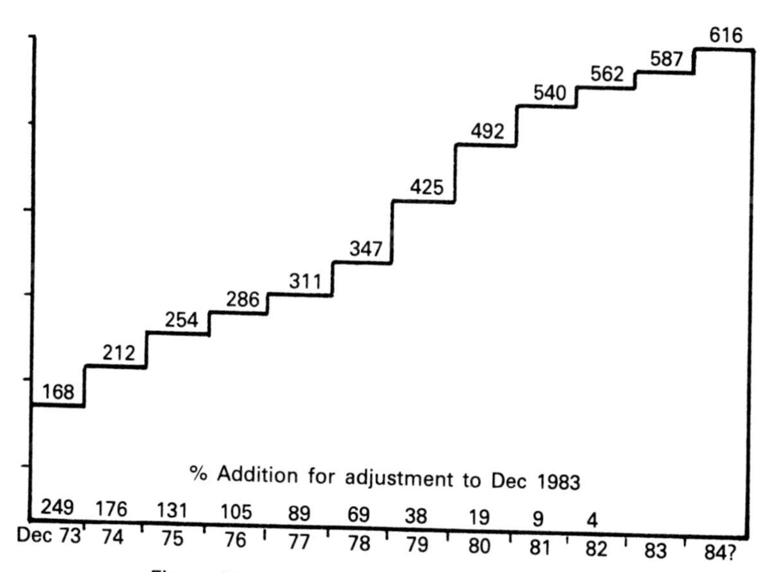

Figure 3.1. General farm building costs, 1973–84

Source: SFBIU, 1984, 'Farm Buildings Cost Guide'

that capital tied up in buildings cannot be liquidated and buildings on their own tend to generate costs rather than income.

In Scotland the advisory services have a computer program (*Ewehouse*) which is used to estimate the profitability and feasibility of inwintering. The program is run on a portable computer terminal which can be plugged into a telephone in the farm office. ADAS offer similar, but more general investment appraisal advice in England and Wales.

The additional returns, costs saved, extra costs, interest rates, inflation, taxation and the likely life expectancy of the building should all be considered when trying to decide whether a project is going to be worthwhile. This chapter is by no means a comprehensive guide to investment appraisal; it is an attempt to provide a step-by-step guide to making a sensible investment decision.

Is a sheep house really necessary?

Generally speaking the agriculture industry in the UK is over-capitalised in relation to future prospects. Where possible, therefore, it is advisable to keep capital expenditure to a minimum. Financial management should be on the same high level as the technical management of our sheep, cows, crops, etc. There are several important questions to ask. Can income be improved without the proposed capital expenditure? What opportunities are there for increasing profit within the present farm system? How does the performance of the farm enterprise compare with that on MLC- or ICI-costed farms? If performance is poor are you vaguely hoping that things will get better if you house the sheep?

Fixed costs account for 70 per cent of the sheep gross margin, therefore high margins are needed for high profits (MLC, 1982). The differences in performance between 'top third' and 'average' producers with MLC-recorded flocks can point to where existing systems might be improved without any capital investment. In 1980 'top third' producers sold sixteen more lambs for each 100 ewes tupped and they kept 6.25 more ewes per hectare. Lowland producers should be aiming for 1.75 lambs sold from each ewe tupped, with a high proportion of these lambs sold fat off grass. Only 15 per cent of recorded flocks produced more than 1.6 lambs per ewe in 1980.

Gross margins were £150 a hectare more on 'top third' farms than on average farms (Table 3.1). Extra lambs accounted for 33 per cent of the difference but stocking rate was the most important single factor at 37 per cent.

Table 3.1. How the top third producers achieved £150 per hectare more than average

	% of extra gross margin
Sale price per lamb	7
Number of lambs reared	33
Flock replacement cost	11
Feed and forage cost	2
Stocking rate	37
Other factors	10

Source: MLC, 1982.

If you are still convinced that building a sheep house is the best way to improve profitability, the next step is to prepare some budgets. All you need is a pocket calculator and the back of an envelope.

Budgeting for change

By how much is the enterprise gross margin going to increase after inwintering your sheep? A partial budget will give you an estimate: add the extra returns and costs saved then compare this total with extra costs and lost returns.

At this stage it is most helpful to list all the changes which might be brought about by inwintering. If for example 600 ewes are to be housed on a predominantly sheep and arable farm, the following things could happen:

- 10 per cent more lambs will be reared.
- Stocking rate will be increased 10 per cent.
- Feed costs will increase.
- Wool clip will be reduced.
- Veterinary costs will be higher.
- Water, electricity, insurance costs will be higher.
- Variable costs of reseeding 8 ha grass will be saved.
- Variable costs of growing 8 ha barley will be incurred.
- There will be 44 tonnes of barley to sell.

Table 3.2. Partial budget: Sheep housing for 600 ewes excluding the cost of the building

	£		£
Extra costs		*Costs saved*	
Extra hay consumption: 0.25 kg/day for 100 days at £70/tonne	1,050	Variable cost of reseeding: 8 ha grassland at £50/ha for 5-year ley	80
Extra veterinary costs: at £0.80p/ewe	480		
Bedding straw: 15 tonnes at £30/tonne	450	*Income gained:* 60 extra lambs at £35 each	2,100
Misc. costs at £0.40/ewe	240	9 tonnes extra hay at £70/tonne	630
Variable cost: of 8 ha barley at £200/ha	1,600	44 tonnes barley at £100/tonne	4,400
Income lost: Wool clip at £0.50/ewe	300		
	4,120		7,210
	Increase in Farm Gross Margin £3,090		

The difference between debits and credits shows that the farm gross margin is likely to rise by £3,090 a year, after the sheep house is built. This is only the first step. The partial budget in Table 3.2 does not include the annual cost of the building, which consists of depreciation, interest, insurance and maintenance.

Calculate the annual cost

An estimate of the cost of the complete project is needed. In the absence of a firm quotation or in the early planning stages there are several sources of farm building cost information:

Advisory services and consultants,
ADAS, 'Cost of buildings handbook',
SFBIU, 'Farm buildings cost guide',
Agricultural journals,
MAFF Standard costs,
Contractors' estimates.

This is also the time to consider the type of house. Choice will depend on the system preferred in the light of relative costs. The pros and cons of the different types of housing are outlined in Chapter 5. Relative costs of the different types of housing are shown in Table 3.3.

If the building is eligible for grant the rates vary and are higher in 'less favoured areas', (Table 3.4). It is important to deduct the grant at the appropriate rate before calculating the annual cost.

Once the cost has been estimated and the bank interest rate is known a judgement can be made on the life expectancy of the building. The length of this 'write-off' period is an important decision. Specialised buildings usually have a shorter write-off period (five years) than more adaptable and flexible structures (ten to fifteen years). The degree of specialisation

Table 3.3. Cost of sheep housing

Type of housing	*Grant eligibility*	*Cost index*
Existing structure	?	
Topless yard	maybe	7–13
Plastic tunnel	yes (frame only)	17
Partly covered yard (new monopitch)	yes	22
DIY portal frame (secondhand material)	yes	27
DIY portal frame (new material)	yes	40
Off the shelf (adequate)	yes	67
Off the shelf (deluxe)	yes	100

Source: Loynes, 1983.

Table 3.4. Rates of grant for sheep housing, 1983-4

	Lowland %	*'Less favoured areas'* %
Agricultural & horticultural grant scheme (AHGS)	22.5	22.5
Agricultural & horticultural development scheme (AHDS)	32.5	37.5

varies with the type of system chosen. Polythene tunnels should probably be written off in five years, but a high, clear-span building can be reasonably written off in fifteen years. Eight years is a fair compromise for most types of sheep house.

The interest rate used should be based on the lending rate of the clearing banks or the Agricultural Mortgage Corporation. Given the capital cost after grant, the write-off period and the interest rate, the annual charge can be worked out quite simply by reference to an amortisation table (Table 3.5).

Example:
Assume that you have decided to build a monopitch sheep house for 600 ewes using new materials. The estimated capital cost before grant is £22 a ewe.

	£
Capital cost at £22 a ewe	13,200
Less: AHGS grant at 22.5%	2,970
Net cost	10,230

If the investment is written off over eight years at 12 per cent, the annual charge will be the net cost multiplied by the appropriate annuity factor from Table 3.5. The annuity factor per £1,000, at 12 per cent over eight years is £202.

$$\text{Annual charge} = \frac{£10,230}{1,000} \times £202$$

$$= £2,067.$$

The estimated cost of insurance and maintenance for the building (£204) should be added to the annual charge above, to give a total amount charge of £2,271. Deduct the annual charge from the increase in farm gross margin (3,090) estimated earlier:

Net profit added to farm business = £3,090–£2,271.

Table 3.5. Annual cost of loan (amortisation table)

	Annual cost (£) per £1,000 borrowed at various interest rates:											
Years	*5%*	*6%*	*7%*	*8%*	*9%*	*10%*	*12%*	*15%*	*20%*	*25%*	*30%*	*40%*
1	1,050	1,060	1,070	1,080	1,090	1,100	1,120	1,150	1,200	1,250	1,300	1,400
2	537	545	553	562	571	578	592	617	658	694	735	820
3	367	374	381	388	395	403	417	439	476	513	552	633
4	282	289	296	302	310	316	330	351	388	424	463	543
5	231	238	244	251	258	264	278	299	334	373	412	493
6	197	204	210	216	223	230	243	265	301	339	379	463
7	173	179	186	192	199	206	219	240	278	316	357	442
8	155	161	168	174	181	188	202	223	261	301	342	429
9	141	148	154	160	167	174	188	210	248	289	332	422
10	130	136	142	149	156	163	177	200	239	280	324	415
11	120	127	134	140	147	154	169	191	231	274	318	412
12	113	119	126	133	140	147	162	185	226	269	313	408
13	106	113	120	127	134	141	156	179	221	265	311	407
14	101	108	114	121	129	136	151	175	217	262	309	405
15	96	103	110	117	124	132	147	171	214	260	308	403
20	80	87	94	102	110	117	134	160	205	253	302	401
40	58	66	75	84	93	102	121	150	200	250	300	400

Thus, the extra income generated by the investment is estimated to exceed extra costs by £819 and the investment appears to be worthwhile.

Allow for tax and inflation

The annual charge is easy to calculate and easy to understand, but it does not take into account the effects of taxation and inflation. Extra net income generated by the building, either because of reduced costs or increased output or both, is subject to income tax. Capital allowances and interest paid on borrowed money reduce this tax liability. Since the March 1984 budget, a farmer is no longer allowed to write off 30 per cent of the cost of new buildings in the first year followed by 10 per cent a year for seven years. From April 1st 1986, only 4 per cent of the cost of a building will be allowed as tax write-off each year. This means that most investments will take twenty-five years to write off.

Inflation affects the annual charge in two ways: it leads to higher interest rates on loans, and reduces the real value of capital allowances that can be offset against income tax. A 4 per cent allowance of £409 on a building costing £10,230 five years after construction has a real value of £285 if the inflation rate has been 7 per cent. This happens because the purchasing power of the tax saving declines by 7 per cent each year. Thus a 30 per cent allowance in the first year followed by 10 per cent for seven years is more favourable than a straight 4 per cent a year for twenty-five years, because the savings are available earlier. When inflation is low and only 4 per cent allowance is available it does not make much difference if you allow for tax and inflation in the annual charge calculation. However, more care is needed when inflation is high and when higher capital allowances are available.

High inflation drives up nominal rates, but real interest rates, the difference between nominal rate and the inflation rate, tend not to vary a great deal. During the last twenty years real interest rates have been around 3–5 per cent most of the time. During the mid 1970s real interest rates were negative for a while, putting borrowers in a very advantageous position.

Real annual capital charge

Real annual charge can be calculated as before except that the real rather than the nominal interest is used.

Example:

Assume as before that you wish to erect a monopitch building for 600 ewes costing £10,230 after grant and it will be written off in eight years. If the inflation rate is 7 per cent and the nominal interest rate is 12 per cent, the real interest rate is 5 per cent. Use Table 3.5 to find the appropriate annuity factor per £1,000 at 5 per cent over eight years.

$$\text{Real annual charge} = \frac{£10{,}230}{1{,}000} \times £155$$

$$= £1{,}586.$$

Add the insurance and maintenance charges as before (£204) and deduct this revised annual charge (£1,790) from the increase in farm gross margin (£3,090). Additional net profit because of the investment will now be £1,300; the proposed investment appears to be an even better proposition when looked at in this way, but there are dangers here. If the overdraft is high and rising even this apparently sound investment could precipitate a cash crisis, especially in the early years. It is also important to look at the expected returns; they may be falling in real terms.

Payback period

How long will it take for the building to pay for itself? This is another useful question to ask although it is a fairly crude yardstick. Generally the shorter the payback period the safer and more worthwhile the investment. Beware of payback periods greater than ten years; the longer the period the greater the risk. Divide the total cost of the new investment by the estimated extra annual income before depreciation on the building.

Example:

$$\text{Payback period} = \frac{\text{Cost of building}}{\text{Extra annual income}}$$

$$= \frac{£10,230}{3,090}$$

$$= 3.31 \text{ years.}$$

It is beyond the scope of a *Sheep Housing Handbook* to go much further into the various methods of investment apprais-al. If it is necessary to submit the investment proposals to more rigorous scrutiny ADAS or a private consultant will work out the 'discounted yield' or 'net present value'. This method is based on the premise that money earned today is worth more than money earned tomorrow, because today's money can be earning interest. The discounted yield will tell you what rate of interest you can afford to pay in order to break even, and if the net present value is positive the investment is probably worthwhile.

Is the proposed investment feasible?

Investment appraisal indicates whether or not the proposed expenditure is likely to be worthwhile in the long run. But if the proposed expenditure resulted in a severe short-term cash flow problem, the project would not be feasible. Sufficient income must be generated to service the loan. Draw up a cash-flow budget in order to find out how much cash is needed and how much cash is earned in each year of the life of the loan.

Example:
Using the information from the first calculation of annual charge and extra annual profit (£819) to prepare a cash-flow budget, see Table 3.6.

Table 3.6. Cash flow for a proposed loan of £10,230 with an initial overdraft of £3,000

Year	*Net annual cash flow* £	*Overdraft repayment* £	*Interest on OD at 12%* £	*Outstanding bank balance* £
0	819	—	—	−3,000
1	819	459	360	−2,541
2	819	514	308	−2,027
3	819	576	243	−1,451
4	819	645	174	−806
5	819	722	97	−84
6	819	809	10	725
7	819	—	—	1,544
8	819	—	—	2,363

The investment should be feasible because the overdraft will be cleared and there will be a positive balance after year six. A shorter repayment period would result in a higher annual charge and a lower net cash flow, the overdraft would then rise during the early stages of loan repayment causing a severe cash-flow problem. Look for an early increase in output as a result of the investment. A long time lag between expenditure and increased earnings can also cause cash problems, irrespective of good long-term prospects.

Profitability and feasibility checklist

- Do you really need a sheep house?
- Can you increase profits without one?

- Do not rely completely on intuition; budget for change.
- Calculate the annual cost of the building.
- Calculate 'payback' period and estimate the risk.
- Will there be enough cash to service the loan?
- Ask someone else to check your assumptions and calculations; they may be more objective.

Reference

LOYNES, J. (1981), 'Putting the right roof over your sheep', *Farm Buildings Digest 16,* 4, 7-8.

Chapter 4

MANAGING THE BUILDING PROJECT

Organising and planning the building

ONCE THE decision has been made to go ahead with investing in the sheep house, the next step is to formulate a clear idea of what you want the building to do. In other words prepare a 'brief' specifying the number of sheep to be housed, the length of the housing period, the feeding system and the type of bedding to be used. Unless you set down these primary requirements at the start, you may not get the building you really want. Once the brief has been prepared it is possible to proceed in any one of several different ways.

The whole project can be handed over to a land agent or other professional consultant. Your consultant will then prepare detailed drawings, organise and supervise contractors. He will take the whole job off your hands for a fee, which is normally 6 per cent of the total cost for large projects and 10 per cent for smaller jobs. Payment of professional fees other than to accountants is often a deterrent to farmers, but if your building knowledge and time are limited, this can be a sound investment.

Package-deal contractors are a useful alternative to the professional consultant if you do not want to hand over the entire job, but do not have the time to organise and supervise construction. After stating your requirements and choosing the best buy, you can leave the contractor to get on with the building. Deciding on the best buy is not always easy as packages are difficult to compare with one another. Design, layout, materials and conditions of sale should all be scrutinised. Take every opportunity to inspect buildings already erected by each contractor. Usually packages are prefabricated and after preparation of the foundations, erection is very fast. Using this type of package, I have had new mono-pitch buildings for 1,000 ewes erected on a prepared site in ten days. If speed is essential, this could be the way to get things moving.

It is possible to prepare your own plans and specifications, invite tenders and supervise construction if you have the time and you are knowledgeable about building construction. Your own sketch plans can be turned into working drawings by a local drafting office, and you can also handle application for grant, planning permission and, if necessary, landlord's permission.

Do-it-yourself construction is the least-cost choice as you are saving the contractor's profit margin, but it does require a great deal of your time and effort together with good basic knowledge of building techniques.

Timing

In most situations a sheep house should be ready for occupation in early January to obtain the full benefits of inwintering in the first year. Work out a realistic timetable for construction. If sheep have to be outwintered as a result of building delays there could be loss of profit, or capital could be tied up in an unproductive building finished in May and unused until the following January. To utilise his labour to the best advantage a building contractor often has to work on a number of jobs simultaneously and sporadically; building time will vary with the type of building. Allow for bad weather in autumn or winter if laying concrete or if building with concrete blocks. Where farm labour is to be used on DIY projects, work can only proceed when the seasonal labour demand on the farm is low, therefore allow for building work to fit in with the overall management of the farm.

Preparing the brief and planning the layout

Try to inspect as many different types of sheep housing as possible, different feeding and construction systems; determine their advantages and disadvantages so as to avoid repeating someone else's mistakes. As an aid to planning ventilation, lighting, water, floor and trough space, penning systems and welfare, these topics are reviewed in the remainder of this chapter.

Siting

Most new buildings are sited in or near to existing farmyards. It is important to consider not only the present requirements of the farm but also the future developments. Try to allow

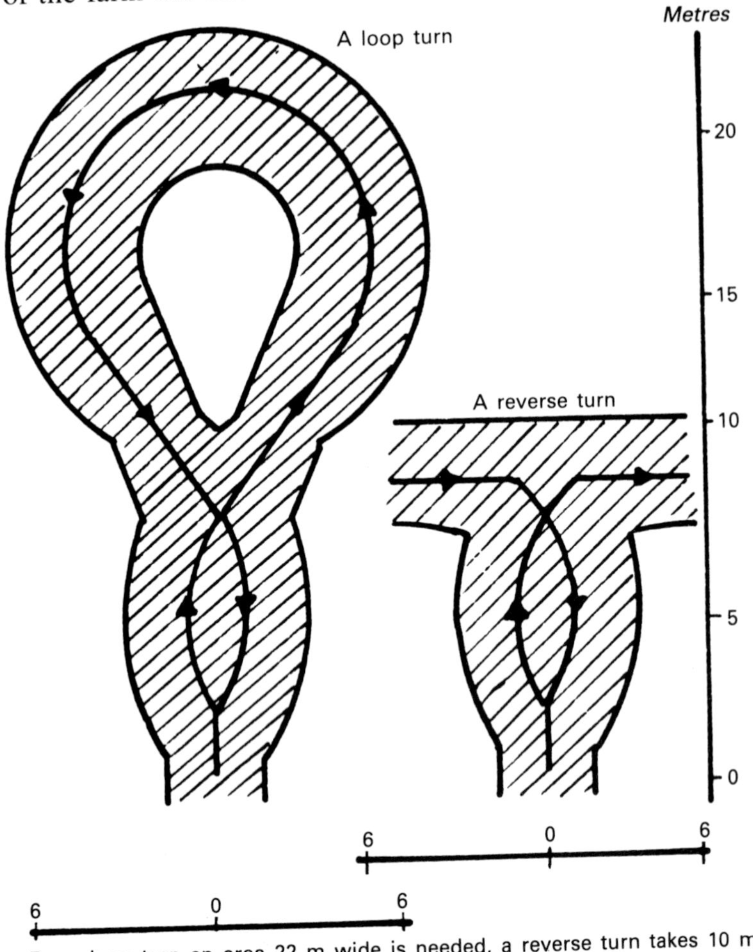

For a loop turn an area 22 m wide is needed, a reverse turn takes 10 m

Figure 4.1. Turning space required for a forage box (overall length 6.4m and overall width 2.4m) towed by a 40 kw tractor

Source: ADAS, 1980, 'Farm Vehicle Movement'

room for expansion and change in farming policy. It is generally felt that it is necessary to live 'on the job' to get the best results in livestock production, therefore how far the sheep are from the shepherd's house should be considered. However try to site the building downwind and below houses to avoid smell and drainage water flowing near houses.

Expansion is a factor often overlooked in the development of farmyards and steadings; allow room to expand in at least one direction. Men, materials and animals should be able to move freely to, from and around the building. Main traffic movements are likely to be the daily feeding of silage or hay, the carting in of bedding and concentrate feeding. Apart from routine footbathing and vaccinations, there is likely to be little movement of sheep to or between buildings until lambing starts. At lambing time grazing should be as close as possible, consistent with clean pasture management for the control of stomach and intestinal worms.

There should be enough space between buildings to allow for maintenance work such as painting, a distance of 2 metres should be sufficient. If tractors and trailers are to pass between buildings though, allow 4 to 5 metres. Forage boxes have wide turning circles (Figure 4.1) and passageways should be designed to take delivery of silage from the forage box (Figure 4.2). A tractor and loader carrying a big bale of

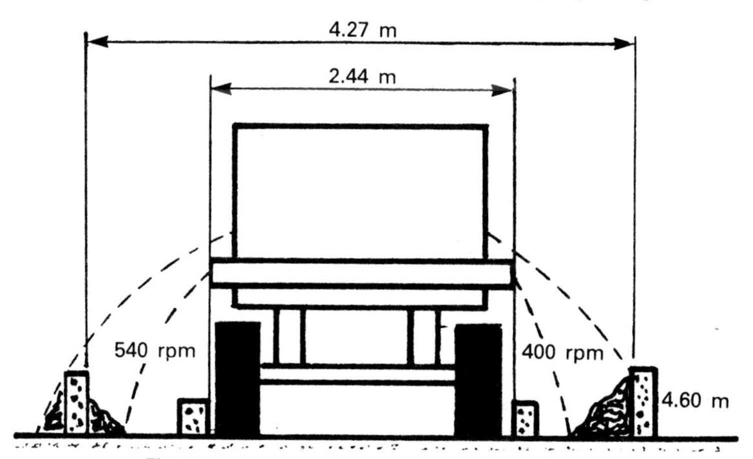

Figure 4.2. Dimensions for forage box feeding
Source: ADAS, 1980, 'Farm Vehicle Movement'

silage needs a similar amount of turning space as a forage box.

When water and electricity services are being installed, a connection point should be left for each so that these services can be easily extended. Plan the layout with safety in mind, so where there is a risk of fire as with hay and straw storage, separate buildings by at least 10 metres and site fuel tanks at least 20 metres from buildings. Ensure that all electrical installation and repair work is done by a competent electrician.

Many new buildings including sheep houses do not require planning permission, because they are regarded as 'permitted developments' under the *Town and Country Planning General Development Order*, 1973. However, in National Parks the planning authorities do have the power to control siting, design, materials and roof colour. Due to the changes wrought on the landscape by modern farming methods and the increased scale of operations there is increasing public demand for planning control on agricultural development, from hedge removal to drainage and building. It is no longer sufficient that the building should only satisfy the farmer's functional requirements. Thoughtful siting and choice of materials can enable buildings to blend with the countryside.

Although function will dictate the siting of most new buildings, every effort should be made to integrate the old and the new. The use of contours, tree planting, cut and fill excavation and appropriate materials can give a sense of unity to a group of buildings for little extra expense. Avoid developments on the skyline, large roof spans and the use of too many different materials.

Siting Checklist

- Allow room for expansion and change.
- Buildings should be downwind from and below houses.
- Allow adequate turning circles for tractors and trailers.
- Ensure access to water and electricity and for maintenance.
- Even if planning permission is not needed, do not create an eyesore on the skyline as appearance is important.

Five basic types of sheep house

By far the commonest reason for housing sheep is the need to get the animals off the land in winter for anything up to four months prior to lambing, with turnout almost immediately after lambing. Where very early (Dec-Jan) lambing flocks are housed, allowance must be made for the ewe and lamb, as housing often continues until the lamb is sold fat out of the house. In the hills inwintering can be an economical substitute for away wintering of ewe lambs. All three systems can be accommodated in any one of the five basic types of house, but floor space, group size, trough space and other requirements inside the house need to be matched to the management system.

There is a wide range of sheep house design and cost. At the low-cost end of the market there are conversions of existing buildings, open 'topless yards' and polythene tunnels. Monopitch, open-fronted houses are about middle of the cost range while some portal-frame buildings for sheep are probably in the luxury class.

In areas of low rainfall open yards may be an adequate means of temporarily removing sheep from the land. However, present experience is limited and anyone taking up this option would be pioneering. Open-fronted yards are a slight improvement but are still limited to drier areas with good drainage underfoot.

Conversions of existing buildings to sheep housing have been carried out successfully where ventilation, trough space and stocking density together with other design requirements have been adequate.

Plastic tunnels for sheep have gained rapidly in popularity because of their low cost, and ewe performance has been as good as that in more substantial, purpose-built sheep houses.

The 'Rosemaund' type of monopitch house has also been widely adopted because of its relatively low cost, ease of construction and suitability for mechanised silage feeding.

None of the above systems are very adaptable for other uses. Only the clear-span portal frame building is truly multipurpose but it is also the most expensive. Even so when sheep are housed for three to four months in the winter other uses are limited.

Plate 9. Ventilation of widespan portal frame building with side windows
Farmers Guardian

Ventilation

There is widespread awareness of the need for good ventilation in sheep houses. If ventilation is poor, humidity increases, water drips from the roof and there is a low rate of air change. When low temperatures are coupled with high humidity and draughts there is loss of heat by evaporation. This combination of conditions can rapidly trigger off respiratory diseases in sheep. Where ventilation is bad there is a rapid build-up of bacteria and viruses in the atmosphere, carbon dioxide levels also increase thus lowering the animal's resistance to disease.

Pneumonia rarely develops when sheep are outwintered. The aim of sheep housing should therefore be to achieve an environment that is reasonably close to that prevailing outside the house without rain, snow, wind and draughts. The air inside the building should change frequently enough to prevent a build-up of harmful disease organisms. This type of environment can be provided by a well-sited, naturally ventilated building.

Intelligent siting can assist natural ventilation as prevailing winds, aspect, shelter, altitude and humidity all have an effect. Choose a site which avoids hollows and if possible run the ridge from north to south. The floor should be free draining to avoid high humidity and condensation problems.

Natural ventilation is possible if air can enter, flow through the house and leave without interference (Figure 4.3). Heat from the sheep raises air temperature above that of the incoming external air and this sets up convection currents inside the building. Layers of hot air then rise by thermal buoyancy or the 'stack effect'. However, it is essential that the building is designed to give minimum ventilation by the stack effect in calm conditions. When the natural ventilation rate is correct, temperature and humidity will not be a problem. It is possible to estimate the minimum inlet and outlet areas needed for the natural ventilation of sheep houses (Bruce, 1982, 1978, 1977, 1975). Given the floor area of the house, the

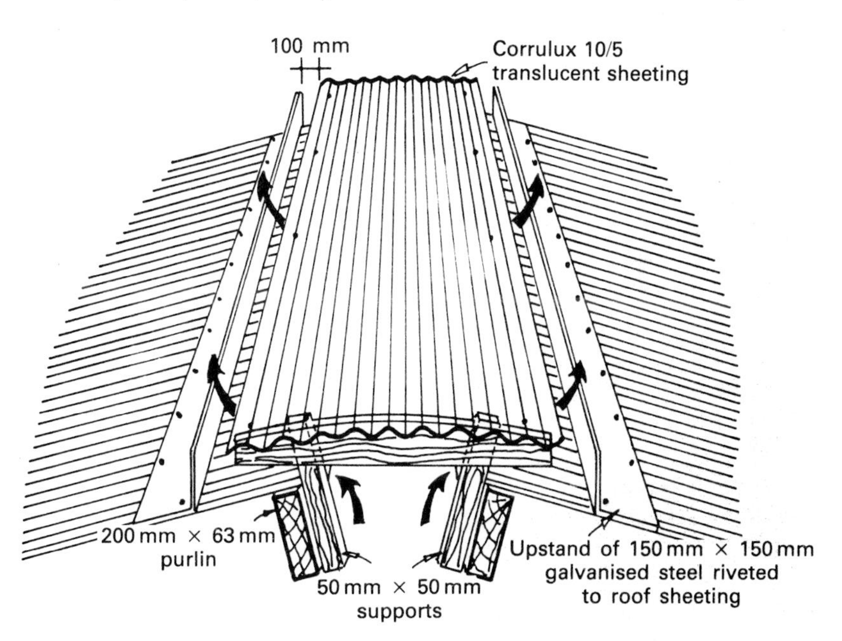

Figure 4.3. An open ridge modified to form a protected open ridge preventing the ingress on rain and snow
Source: West of Scotland College of Agriculture, 1982

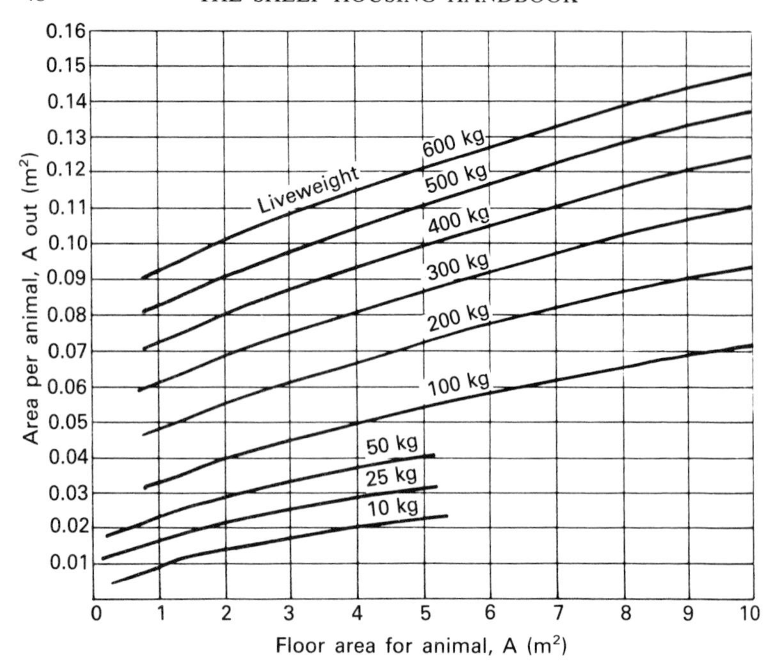

Figure 4.4. Calculation of open ridge area. Outlet area for a height difference of one metre between the eaves

Source: Bruce, J.M., 1982

body weight of the sheep and the pitch angle of the roof, it is possible to estimate the size of open ridge needed for ventilation (Figures 4.4 and 4.5).

To prevent draughts, some solid but light wall cladding should be used at floor level extending at least to 1.2 metres above the final height of any straw bedding.

How much floor space?

Sheep vary in size with breed, age and stage of pregnancy, therefore recommended indoor stocking rates also vary. Err on the side of caution when planning the building as it is better to start off with too much space than too little. Stocking density can be increased with experience. If the flock is to be shorn at housing, it is possible to reduce the recommended space allowances by up to ten per cent.

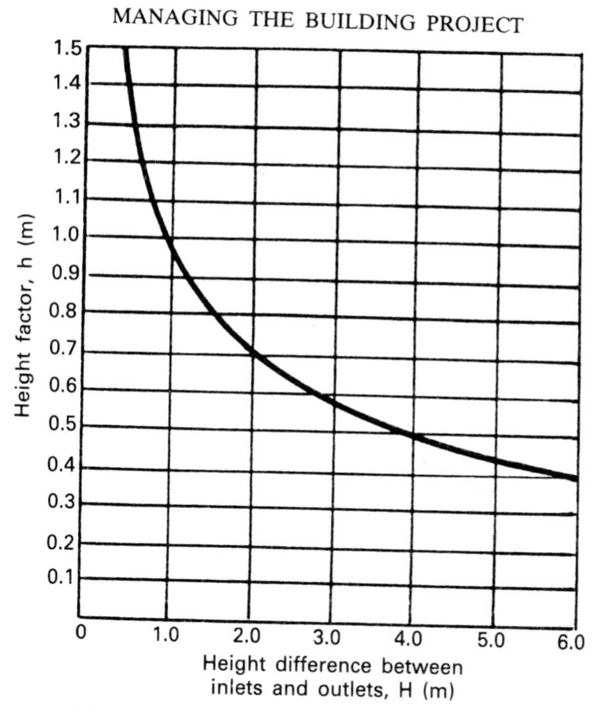

Figure 4.5. Height correction factor for differences between eaves and ridge greater than one metre
Source: Bruce, J.M., 1982

Table 4.1. Recommended floor space for sheep

Type and body weight	Floor space m²	
	Slats	*Straw*
Large ewe 60-90 kg in lamb	0.9–1.1	1.2–1.4
Large ewe 60-90 kg with lambs	1.2–1.7	1.4–1.8
Small ewe 45-60 kg in lamb	0.7–0.9	1.0–1.3
Small ewe 45-60 kg with lambs	1.0–1.4	1.3–1.7
Hoggs 32 kg	0.5–0.7	0.7–0.9
Lamb creep (2 weeks)	NA*	0.15
Lamb creep (6 weeks)	NA*	0.4
Ewe plus weaning lambs in pen		2.2
Lambing pen		1.5

* It is not advisable to keep young lambs on slats; many injuries can occur due to feet being smaller than the slatted gap.
Source: Farm Buildings Information Centre (1983), *Housing Sheep*.

Pen layout

Opinions vary as to how many sheep to keep in one group within a house. Groups have to be large enough to minimise the cost of internal divisions, but not so large that they are difficult to control or to observe. I have seen over a thousand high-performing ewes in one undivided sheep house in Spain, but in practice under UK conditions groups of from twenty-five to fifty appear to be the best compromise between economy and management requirements. However, where unmated ewe lambs are inwintered as an alternative to away wintering up to 100 can be kept in one pen quite successfully.

Divisions between pens should be made of the most economical material available; these may be railed fences or sheeted with corrugated iron, exterior plywood or plastic netting (Netlon). It is useful to be able to raise the pen divisions as the depth of bedding increases later in the winter. Walk-through hay and concentrate troughs can also be used as pen divisions where hay or straw are fed by hand.

Straw bedding or slatted floors?

Traditionally straw littered yards have provided a warm, clean and dry environment for livestock, but during the mid 1950s there was a move towards slatted flooring in the pig industry. High stocking densities, lower labour requirements and cleaner, drier pens were the claims made for slatted pig fattening pens. As it is vitally important for housed sheep to have a clean, dry floor, the use of slats in sheep housing was one of the first investigations carried out by the Sheep Unit at the NAC, Stoneleigh. Timber, metal and concrete slats were compared and, as a result, it is now possible to give a specification for slatted sheep house flooring. As long as the material is strong enough timber, metal or wire mesh can be used. The floor should consist of slatted panels for ease of access to the dung chamber underneath. The slats themselves can vary from 25 mm with a 15 mm gap, up to 100 mm with a 20 mm gap; the latter example being for larger sheep. Slats have many advantages: footrot problems are minimised, fleeces are cleaner, stocking density is higher, less labour is needed when bedding is eliminated, and a drier atmosphere can be maintained. But slats do have one big disadvantage which tends to outweigh all of their advantages on most farms: they are very expensive. Slats can easily double the cost of a building.

Where straw bedding is used the floor material should be free-draining — brick rubble, stone or gravel can be used to achieve this requirement. Although stock density is less than on slats, cost per ewe housed is much less.

Straw usage varies with breed, size of sheep and stocking density: a reasonable estimate would be two bales per ewe during an inwintering period of 100 days. Although foot problems may be greater, more labour may be needed or straw may even have to be purchased, straw bedding is likely to be cheaper than slats in the long term. Concrete floors are expensive, cold and damp but it may be advisable to have at least a small section of the floor concreted to provide an easily cleaned surface for the lambing pens. Do not use sawdust to bed lambing ewes as it contaminates the fleece, balls up under their feet and they will not lick their newborn lambs when they are covered in sawdust. Sand is a good alternative to straw if it is readily available.

How much trough space?

It is absolutely essential that housed sheep have enough trough space. The amount of space each ewe needs at the trough depends upon body weight, stage of pregnancy and

Table 4.2. Trough space for different breeds

Breed	*Weight at mating (kg)*	*Length of trough per ewe* Concentrates (mm)	*Forage (mm)*
Lincoln	91	500	225
Scotch halfbred	80		
N. Country Cheviot	73		
Mule	65		
Clun Forest	60	450	200
Scottish Blackface	54		
Lleyn	53		
Swaledale	48		
Lambs	45	400	175
Lambs	23	300	125

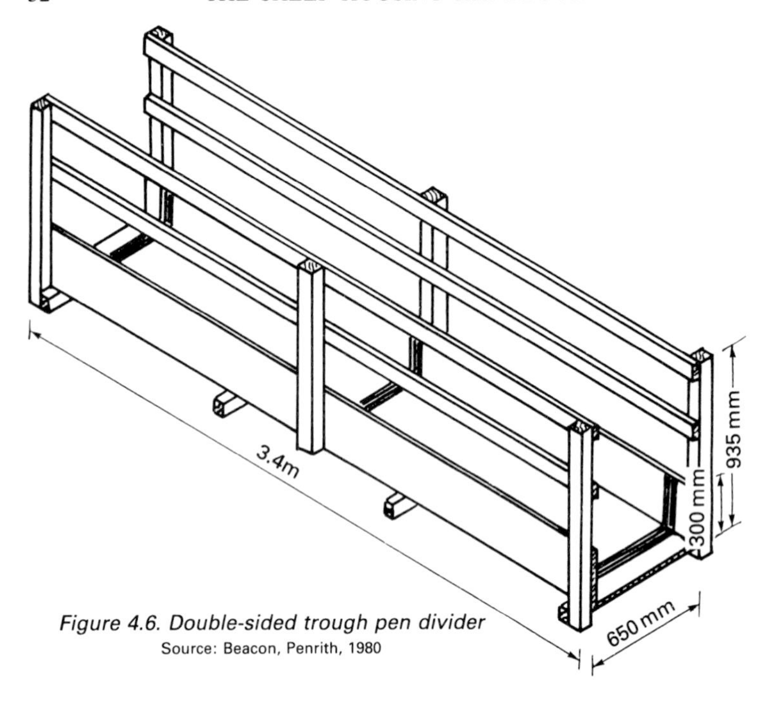

Figure 4.6. Double-sided trough pen divider
Source: Beacon, Penrith, 1980

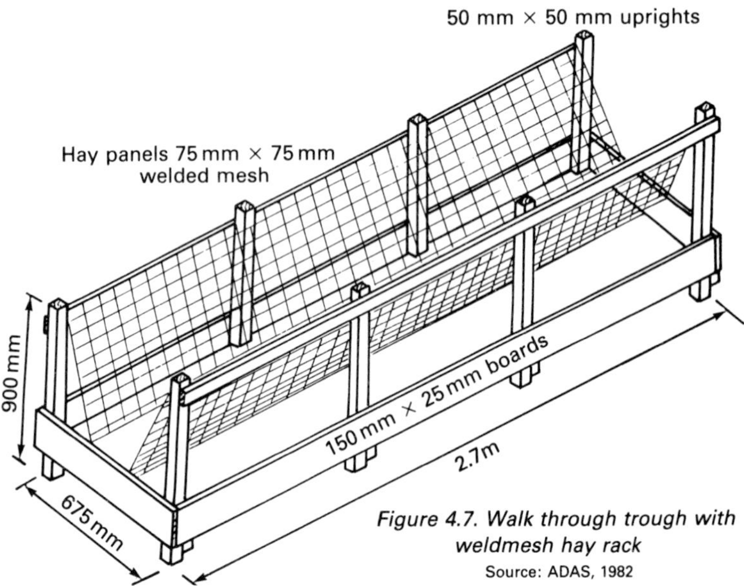

Figure 4.7. Walk through trough with weldmesh hay rack
Source: ADAS, 1982

feeding system. A range of trough spaces that have been found adequate in practice for different breeds is given in Table 4.2. As with floor space, trough space can be reduced by 10 per cent if the ewes are shorn at housing.

Due to the relatively long trough space needed for concentrate feeding in late pregnancy, sheep houses and pens tend to be long and narrow. The trough or feed fence can form two sides of a feed passage as in Figures 4.6 and 4.7. This is the normal layout for controlled feeding. Where ad lib hay or silage is fed, the pens can be less elongated or even square. Pen divisions can be replaced by feed troughs for hand feeding of hay. Where existing buildings are to be converted, it is possible to give the ewes enough trough space for concentrates by using the ordinary feeding troughs brought in from the field or even by feeding ewe cobs on the floor. Similarly, bunkers for big bale silage can reduce the need for long pens and long runs of expensive roofing. The ideal dimensions for feeding troughs are given in Table 4.3.

Water

A supply of clean, fresh drinking water must be available at all times. It is thought that sheep prefer to drink from open troughs, rather than bowls, and that hill sheep prefer a supply of running water. Self-filling water bowls allow a greater degree of flexibility in pen design and tend to be cleaner than open troughs. If open troughs are preferred, they should be raised off the floor on a plinth with an overflow pipe to the outside of the building. A trough measuring 0.6 m × 0.3 m is recommended for a pen of 80 ewes and one large trough can

Table 4.3. Sheep trough dimensions

	Size range (mm)	
	Ewes	*Hoggs and feeder lambs*
Throat height	400 max	350 max
Height from floor of trough to throat	150–200	—
Opening above throat height for access to trough	200–300	—

Source: Robinson, 1982.

Plate 10. Sheep feeders in the feed and load position at Rosemaund, Preston Wynne, Hereford *Farmers Guardian*

Plate 11. A combined hay and concentrate trough used as a pen division *Farmers Guardian*

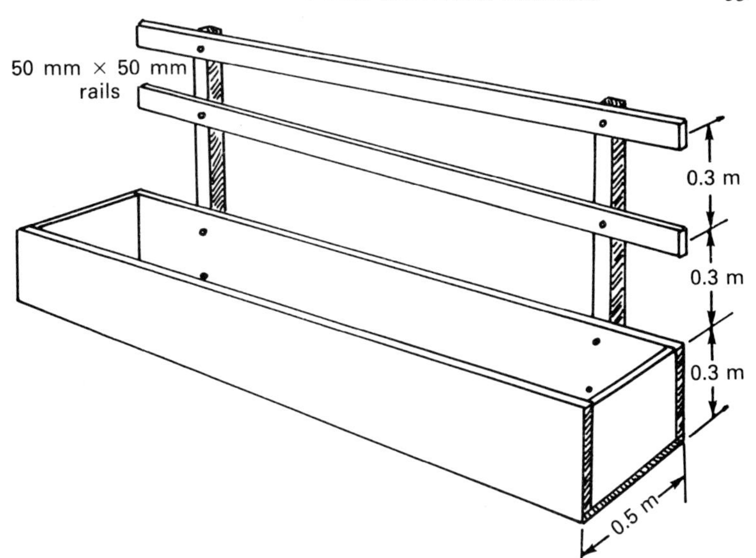

Figure 4.8. Rosemaund combined silage trough and pen front

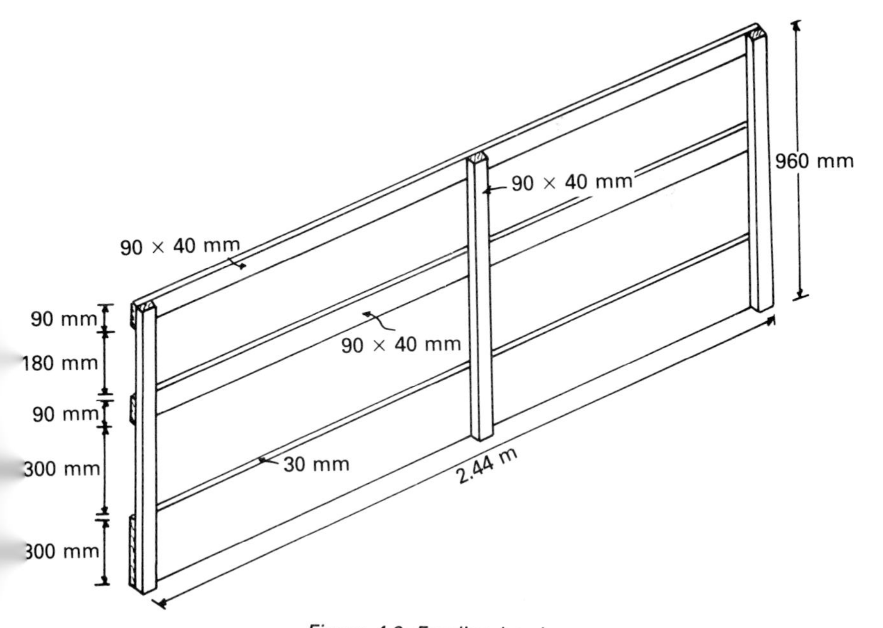

Figure 4.9. Feeding barrier

Plate 12. Fresh drinking water piped the length of a sheep house
Farmers Guardian

serve two adjacent pens. I have seen running water supplied very successfully to hill sheep by diverting part of the flow of a stream through a glazed round tile trough along the back of the house.

Make sure that the water supply to the troughs is protected from frost, because if ewes are deprived of water in late pregnancy, it can precipitate twin-lamb disease. Lag the supply pipes and then protect the lagging with wire netting or wood. Troughs can also be protected by low-voltage heating elements.

Lighting

The code of recommendations for the welfare of sheep, *Agriculture (Miscellaneous provision) 1968 Part 1*, states that: 'throughout the hours of daylight the level of indoor lighting, natural or artificial, should normally be such that all housed sheep can be seen clearly. In addition, adequate lighting should be available for inspection at any time'. As purpose-built sheep houses tend to be open-sided, there is usually a considerable amount of natural light available. Where buildings have been converted however, this may not be the case and it may be necessary to put timber-framed roof lights into tiled roofs or translucent sheets into asbestos or corrugated iron roofs.

Artificial lighting is essential at lambing time and for nightly inspections at other times. Four watts per square metre is the recommended minimum level of illumination. Although fluorescent lighting is more expensive to install, running costs are about a third of those for ordinary light bulbs. All electrical installations should be out of reach of the sheep.

References

Bruce, J.M. (1983), 'Natural ventilation of sheep buildings', *Housing Sheep*, FBIC Stoneleigh.

——, (1981), 'Design method of natural ventilation of cattle buildings using raised roof sheets', *Farm Buildings Progress (66)*, 17-21.

——, (1978), 'Natural ventilation through a vertical opening', *Farm Building Progress (53)*, 9-10.

——, (1975), 'Natural ventilation of cattle buildings using thermal buoyancy', *Farm Buildings Progress (42)*, 17-20.

Cermak, J.P. (1974), 'Farm buildings, the planning approach', SFBIU.

Court, K. (1978), 'Slatted floor trial at NAC Sheep Unit', *Farm Buildings Digest 13*, 2, 19-20.

Hives, J.K. (1983), 'Think before you build', *ADAS leaflet 835*.

MAFF (1977), 'Codes of recommendations for the welfare of livestock, Code No. 5, Sheep', *Agriculture (Miscellaneous provisions) Act 1968 Part 1*.

Robertson, A.M. (1978), 'Sheep Housing', *Farm Buildings Progress (51)*, 1-4.

Robinson, T.W. (1982), 'Designing for the confinement of sheep', *Farm Buildings Progress (69)*, 9-14.

Wright, H.J. (1975), 'How to organise a farm building contract', SFBIU.

Chapter 5

TYPES OF SHEEP HOUSING

THE COST of a new building varies considerably with the design, the materials used and the method of construction. At the cheapest end of the scale conversion of existing buildings may only involve the provision of feed troughs and pen divisions, in a case where the ventilation is adequate and no structural alteration is needed. New buildings on the other hand can range from £5 per square metre for roofless yards, to £10 per square metre for polythene tunnels and as high as £45 per square metre. In the latter case the building would probably have a steel or concrete frame and would be erected by a contractor as part of a package deal.

Initial capital cost should not be the only consideration, cost effectiveness is just as important. Cost effectiveness is difficult to predict but the realistic life of the building, annual maintenance cost and overall efficiency can be used as yardsticks when comparing different housing systems. Although the initial capital cost of a polythene tunnel house may be as low as £10 per square metre, with a life of five years and no maintenance cost, its annual cost will be £2 per square metre a year. On the other hand, a steel-framed building roofed with asbestos or steel sheeting might cost £35 per square metre initially, but with a working life of forty years and £10 per square metre maintenance charge, in that time the annual cost would be as little as £1 per square metre. What appears to be the cheapest initially may not be the best buy in the long run.

Do-it-yourself

DIY building can often be used to keep costs down, as the contractor's profit margin can be saved and farm labour is cheaper than that employed by a builder. Further savings can

Plate 13. Redundant buildings with good ventilation make cost effective sheep houses

Plate 14. Former bull pens converted to sheep housing *Farmers Guardian*

Plate 15. Sheep housed in a converted turkey shed *Farmers Guardian*

Plate 16. Dutch barns convert readily to sheep housing

be obtained by shopping around for the best buy in new materials or by using secondhand timber and sheeting. DIY is not necessarily suitable though for every person or situation.

Home-grown chestnut poles, secondhand galvanised sheets and farm labour were used to construct the sheep housing in Plate 23. The very low cost, approximately £0.75 per square metre in 1980, has enabled the owner to stock the houses relatively lightly at 2 m^2 a ewe, which can be very advantageous at lambing time. The buildings are not attractive but are well screened from the house and the road by trees. This low-cost housing has made it possible to keep 690 ewes on 48 hectares of grassland all the year round.

Although it can be the most effective and cheapest way of building for those with the skills, if the skills, time and labour are not available a good case can be made for buying a package-deal building direct from the manufacturer. Suppliers of farm buildings who are also members of the Agricultural Construction Industry Federation offer buildings to a minimum standard of design and construction. British Standard 5502 is the current guideline for quality buildings.

Converting existing buildings

One of the most effective ways to gain the economic advantages of inwintering is to convert an existing structure which is no longer needed for its original purpose. Disused cattle yards, dutch barns, grain stores, in fact any building which will provide good ventilation, a dry bed and sufficient trough space can be adapted for sheep. The dutch barn in Figure 5.2 has been covered with netlon on the weather side and is as effective as any purpose-built sheep house. In fact the original conversion was so effective that two lean-to extensions were built in succeeding years using secondhand timber and farm labour.

At the 'Scotsheep' event in 1981 a great deal of interest was generated by a low-cost sheep house designed at the North of Scotland College of Agriculture. A frame of an old Nissen hut was the basis of the house and several different conversions were demonstrated. The hut frame could either be attached to sleepers or the sections could be buried in the ground for stability (Figure 5.1). Walls were clad with polythene stretched over the frame and ventilation was through an open ridge. In 1981 the shed cost £4.00 a ewe to convert.

Roofless yards

A roof and its supports are often the most expensive part of a building and attempts have been made to manage without a roof altogether where sheep and cattle are confined. These attempts have had varying degrees of success. The Sheep Demonstration Unit at Stoneleigh and the Edinburgh University Farm at Boghall, 2,000 ft up in the Pentland Hills, have both tried roofless wintering compounds for sheep.

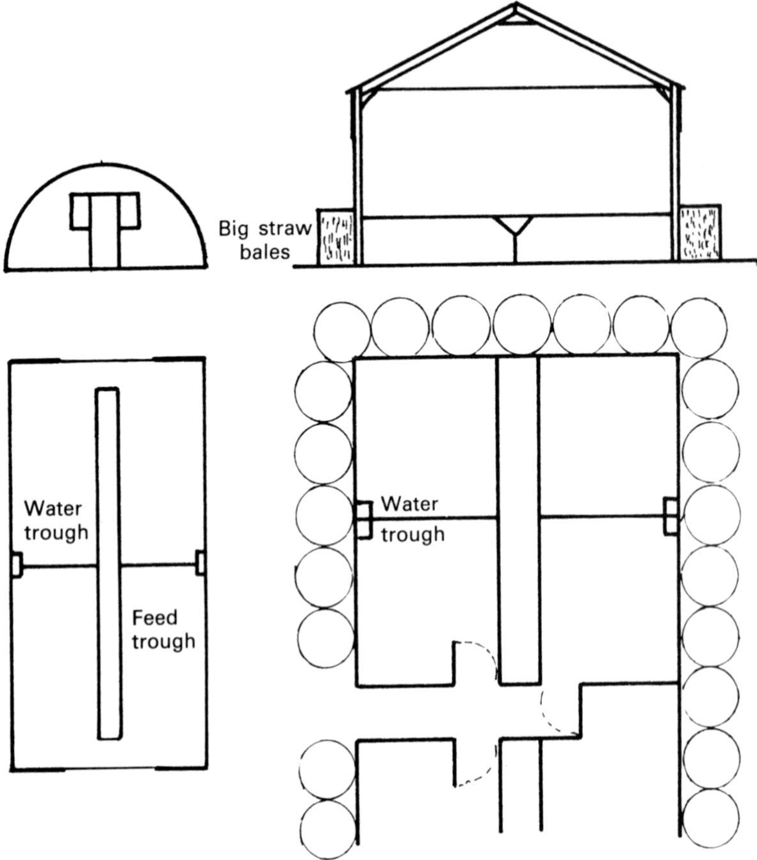

Figure 5.1. Nissen hut conversion *Figure 5.2. Dutch barn conversion*

Source: North of Scotland College of Agriculture, 1982

At Boghall the original roofless yard consisted of long, 3 metre wide pens separated by a 3.3 metre wide feeding passage (Figure 5.3). As the main function of a roof is to keep both the animals and their bedding dry, dry bedding had to be achieved by provision of a well-drained floor. Hardcore 150 mm deep with an aggregate size down to 12 mm was laid on 'Terram' fabric placed on the existing soil. Perforated drainage pipes were laid at 4 metre intervals in trenches backfilled with gravel. After the first year it was found necessary to change the floor material to a single size aggregate of 50 mm. The dust and smaller aggregate particles in the first floor tended to impede drainage.

The perimeter walls were made of 'Tensar' snow fencing, 1 metre high, strained between sleeper corner posts with intermediate posts every 2.5 metres. Tensar has a 50 per cent void area and is a very effective windbreak. The pens were designed to provide Scotch halfbred ewes with a floor area of 1.4 square metres and 50 cm trough space. When used for fattening lambs or for hogg wintering the pens were stocked

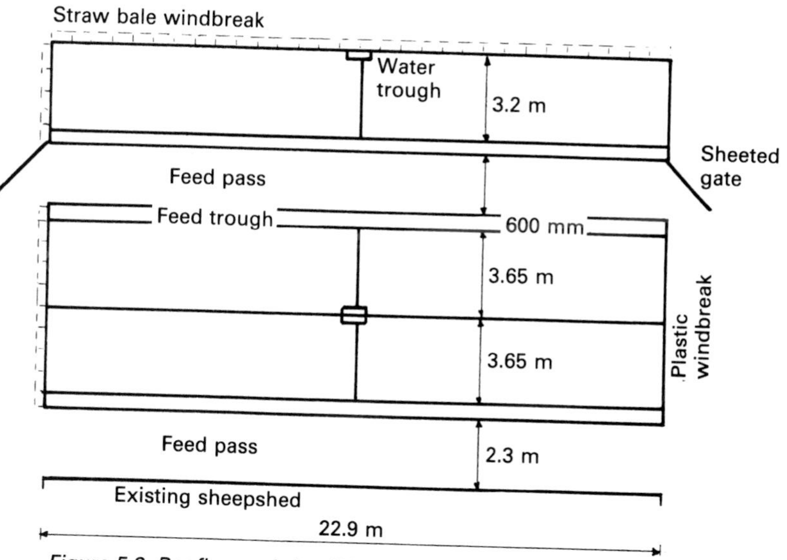

Figure 5.3. Roofless unit for 150 sheep, East of Scotland College of Agriculture

Source: Watson, G.A.L., 1982

at 1 square metre per animal with 30–40 cm trough space.

Silage, hay or chopped swedes could be fed in the simple box troughs lining the passage ways; these also acted as a retaining wall. Water was provided in troughs rather than in self-fill bowls to reduce the risk of freezing. Frozen water supplies could be a problem with this unprotected yard. Incoming pipes should be buried at least 90 cm deep and all above-ground pipes should be well insulated. Ball valve joints can freeze in cold weather too; if thawing these out with kettles of boiling water is too arduous, heating elements are available to prevent freezing.

Foot problems were not as bad as expected, but the sheep became very dirty in wet weather and more straw was needed. In dry cold weather the sheep stayed clean. Twice as much straw was used in the roofless compound compared with roofed accommodation; about 24 kg per lamb during a ten-week period.

The roofless compound was cheap; only £4.50 a ewe in 1981, but several drawbacks to this system became apparent during the winter of 1981–82:

Plate 17. Ewes wintered in roofless yard at Redesdale EHF

- Roofless pens should only be used for a relatively short period before lambing, as the bedding becomes very wet and the sheep very dirty.
- Despite what seemed to be the good drainage characteristics of the hardcore floor, the straw on top of the hardcore acted like a sponge, that is, it absorbed and held water which did not then drain through the floor.
- Roofless pens should be restricted to areas of low winter rainfall.

Plate 18. Concentrate feeding troughs in a roofless yard are pivoted to keep them clean and dry after use — Caverton Hillhead, Kelso

- In 1983 the pens were provided with a roof to keep the bedding, sheep and shepherd dry.

The problems of wet bedding could be overcome by using slatted floors (Watson, 1982). Slats of steel mesh, softwood or hardwood could be used, the latter being more durable and more expensive. But even softwood slats are expensive, and a roof could be provided for the cost of a softwood slatted floor (Table 5.1).

Table 5.1. Building costs 1982 — (300 ewes)

	£		£
Roofed bedded pen	9,000	Roofed slatted pen	13,500
Roofless bedded pen	4,500	Roofless slatted pen	9,500
Saving	4,500	Saving	4,500

Source: Watson, 1982.

A pen similar to that on the Edinburgh University farm at Boghall was erected at the NAC Sheep Demonstration Unit, Stoneleigh in 1978. The pen was used during the last half of the winter for the four-and-a-half weeks prior to lambing.

Aggregate 150 mm deep was used for the pen floors, but it was laid to a greater depth in the feed passage where a forage box passed twice a day. Windbreak walls were constructed of ICI 'Paraweb' 1.4 metres high. During the first winter when the ewes were confined for only four-and-a-half weeks there were no health problems and the exercise was considered to be successful. However, since then experience has shown that in a wet winter, wet bedding leads to intractable foot problems and the roofless confinement has been discontinued.

There are farms where short-term roofless confinement in straw yards has a place. At Caverton Hillhead, near Kelso a roofless compound is constructed every year with big straw bales as a windbreak and stock-proof barrier on a south-facing, sloping site (Figure 5.4). Lambing begins at the end of March and the ewes are only penned in the yard for four weeks. At the end of the winter penning for only a relatively short period avoids most of the snow which could drift inside the yard. But more important, having the ewes off the land at the end of February enables an early start to spring ploughing of old grass and establishment of spring cereals. The floor consists of 150 mm deep hardcore as at Boghall.

Round pebbles, 25 mm diameter, have been successfully used as an alternative to hardcore when covered with straw for the floor of a roofless pen (Robinson & Money, 1982). A layer of pebbles 150 mm deep was laid over drainage channels 600 mm deep and 1.5 m apart. The channels were then

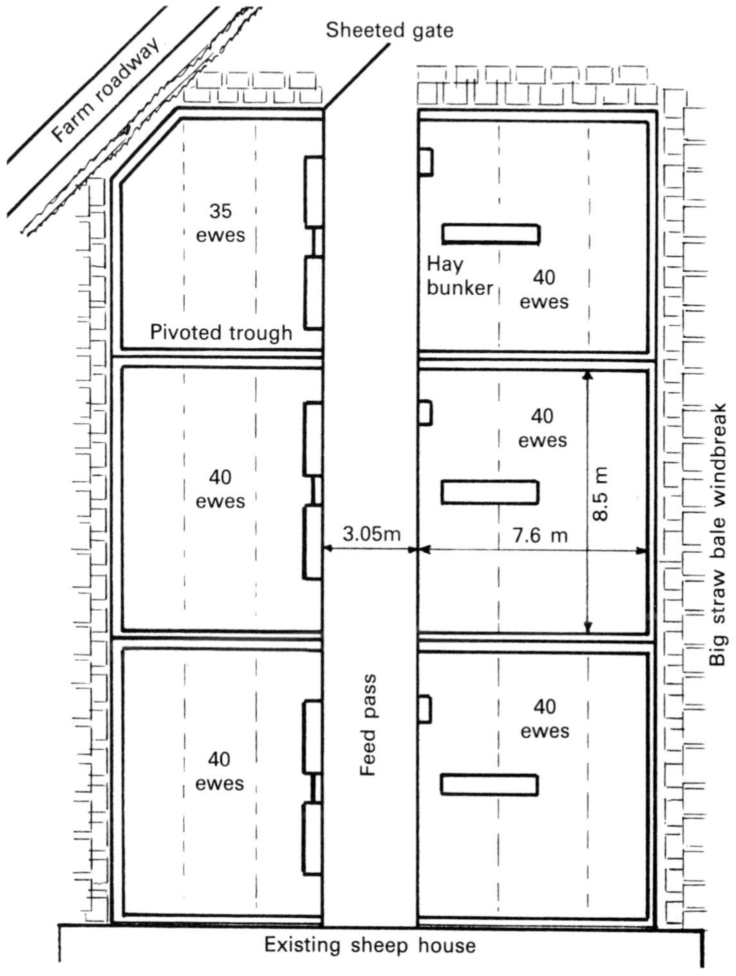

Figure 5.4. Roofless unit for 235 sheep
Source: Watson, G.A.L., 1982

backfilled with 75 mm graded stone. Rainwater and rotation of the pebbles moves faeces down the floor profile; there is no water absorbing straw. The perimeter fencing was made of 'Tensar' and the pens were stocked at 0.83 square metres a ewe. At lambing the ewes were moved to covered accommodation.

In the British winter, roofless sheep pens with straw bedding do not seem at all satisfactory but some compromise is possible. A roofed lying area combined with a roofless exercise yard seems to work quite well and has been quite widely adopted where old buildings have been adapted.

Polythene tunnels

After conversion of existing buildings polythene tunnels are one of the cheapest types of farm building. Originally developed for use in horticulture they have now been widely adopted for sheep housing. On some farms, particularly in exposed upland areas plastic tunnels have sometimes failed to stand up to winter storms, gales and blizzards. During the winter of 1981–82 a number of polythene tunnel sheep houses caved in after heavy falls of snow. At the same time some tunnels exposed to similar conditions did not suffer structural damage. The reasons for this difference in ability to withstand high wind speed and snow loading seem to be due to a number of factors:

- The diameter and gauge of metal used in the main support hoops; the larger and thicker the better.
- Distance between the main support hoops.
- Method of fixing hoops to the ground.
- Siting of the tunnel in relation to the prevailing wind.
- Area and location of ventilators within the tunnel.

Until 1976, when many polythene tunnels were blown down in the winter gales, designers and manufacturers based their designs mainly on guesswork or trial and error. However, at the National Institute of Agricultural Engineering (NIAE) engineers have been able to produce a model of the effect of wind on plastic tunnels used in horticulture and livestock housing. The problems that winds of varying speeds impose on plastic tunnels can now be predicted and the information is available to manufacturers.

If the design is right plastic tunnels will stand up to very extreme conditions. Before buying a plastic tunnel to house sheep check with the manufacturer that NIAE specifications have been incorporated into the design. This is especially important if the tunnel is to be used in an exposed upland or coastal area.

One of the first polythene tunnels to be used for sheep housing was erected at the NAC Sheep Demonstration Unit, Stoneleigh. Performance of the sheep and climatic conditions within the house were closely monitored by ADAS during the winter of 1979–80 and 1980–81 (Cornwell-Smith, 1981). During the second winter, observations were carried out by ADAS staff in other parts of the country on similar types of polythene house. The results for different sites and climatic conditions were then compared (Table 5.2).

Table 5.2. Temperature records: polythene tunnel housing

Site	*No. sheep housed*	*Dates recorded*		*Average temperature °C* *Inside*		*Outside*		*Lift over Outside*	
				max	*min*	*max*	*min*	*max*	*min*
NAC	120 Scotch halfbred	20/12	18/3	9.1	2.1	7.9	1.1	1.2	1.0
West Midlands	120 Mules	13/1	10/3	10.6	3.0	7.7	2.4	2.9	0.6
Kent	114 Romneys	2/3	3/5	14.1	6.7	12.3	5.5	1.8	1.2

Source: Cornwell-Smith, 1981.

Polythene tunnels used in horticulture were not suitable for sheep housing until the ventilation was improved. At each side of the house the polythene wall was replaced with 'Tensar' windbreaker material to a height of 1 metre. Good inlet ventilation was provided at sheep level and windspeed inside the house was reduced. Outlet ventilation was obtained at the top of the house by holding open a gap of about 30–40 cm where the polythene sheets overlap. Both gable ends of the house were covered with 'Tensar' netting to reduce windspeed and to provide some shelter.

Temperature, windspeed and humidity were all monitored at different points inside and outside the building. Average temperature inside the house was about 1–2°C above the outside air temperature. At low windspeeds the 'Tensar' netting walls tended to reduce windspeed inside the building, but this was only for short periods and the sheep were not

Plate 19. Silage feeding in a polythene tunnel. The troughs were made cheaply from onduline roofing sheets

Plate 20. Polyester reinforced PVC tunnels on an exposed site in Northumberland *Farmers Guardian*

affected. Humidity was found to be almost always the same inside and out; only when silage was fed did the internal humidity rise above that outside.

After the first year's experience with polythene housing at Stoneleigh the idea was adopted on many other sheep farms. Four Polypen houses were erected on the Leeds Castle Estate in Kent during 1981. Each tunnel houses 200 ewes with floor space allowance of 1.25 square metres for each ewe. Since the structure was erected there have been no health problems and ventilation has been found to be perfect.

During the winter of 1981 the conditions in the two other plastic tunnel houses were monitored by ADAS (Table 5.2). Results were very similar to those obtained at Stoneleigh, the 'Tensar' netting walls reduced windspeed by about 75 per cent. There were no discernible draughts at sheep level even with a gale blowing. Ventilation was good even on still days. The Stoneleigh ewes were shorn in January, and even when the temperature fell to freezing point the ewes did not seem to be uncomfortable. In April when the temperatures reached

Plate 21. Pens and troughs laid out for hay feeding in a polythene tunnel

16°C the shorn sheep were less stressed than the unshorn.

Careful preparation of the site is essential if the pen floors are to be well drained. The Leeds Castle houses were erected on a wet site and experience has shown that a floor with a convex profile — high at the centre of the house and sloping to the foot of the walls — is the best way to keep water out. One major disadvantage of polythene tunnels with Tensar walls is that water tends to run down the roof and into the pens through the plastic mesh wetting the bedding. This can be overcome to some extent by fixing plastic gutters along the bottom edge of each polythene sheet to carry water away from the wall. A french drain along the foot of each wall also helps to keep the bedding dry.

The polythene skin usually lasts for three years but can last up to five years. Extending the life of the skin depends on

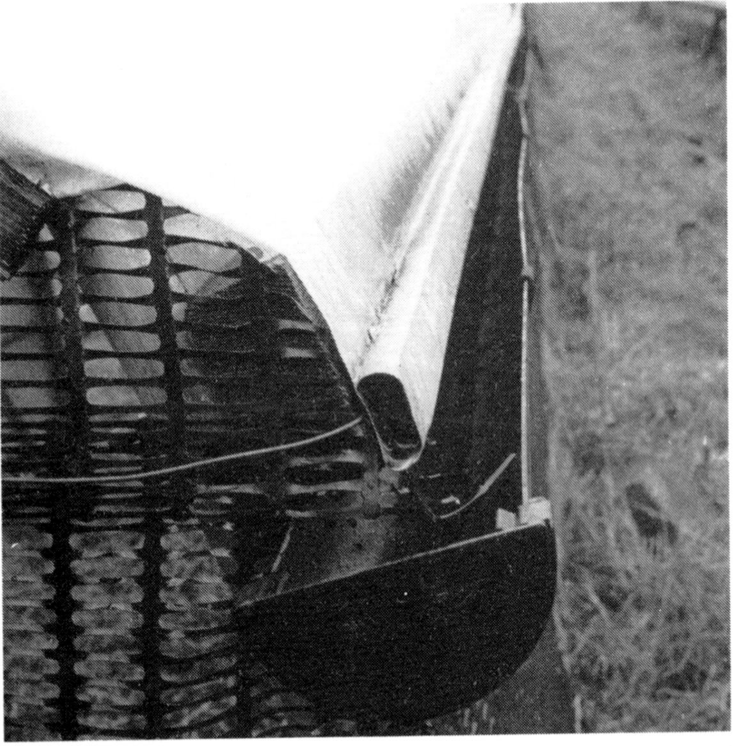

Plate 22. Plastic guttering helps to keep bedding dry in a polythene-clad tunnel

good maintenance. The secret is to get the plastic film as tight as possible when the structure is first erected and then to tighten it each year to prevent chafing in the wind. Polythene can become quite slack in the summer and should be tightened before the autumn gales. Good siting also helps to prolong the life of the polythene; avoid exposing the side of the house to the full force of the prevailing wind and choose as sheltered a site as possible.

An alternative to polythene sheeting is polyester-reinforced PVC. Although more expensive, this material has a guaranteed serviceable life of at least five years. Inlet vents are included in the side walls of the polyester-reinforced material, so water is less likely to enter these walls than through the Tensar netting. But the polyester-reinforced material is more expensive and increases the cost per ewe, despite the longer life of the cladding (Table 5.3).

Table 5.3. Cost per ewe housed in polythene and polyester-reinforced PVC tunnels

	Polythene cladding	*Polyester-reinforced PVC*
Dimensions (m)	30 × 9	30 × 8.5
Area (m^2)	270	255
Ewes housed 1.25 m^2 each	190	180
Cost (£/m^2)	7.50	12.00
Cost (£/ewe)	10.65	17.00

Polyester-reinforced tunnels stand up to exposed conditions much better than the cheaper polythene-clad tunnels. For very exposed sites the manufacturer doubles the number of support hoops which are then anchored in concrete. A polyester-reinforced tunnel with extra support hoops was erected at the West of Scotland College of Agriculture hill farm at Crianlarich in 1983. This house survived a very heavy snow loading in January 1984 when a polythene-clad house adjacent to it collapsed. In England and Wales grant aid is available for the framework but these structures are not eligible for grant aid in Scotland.

If hay rather than silage is to be fed in a polythene tunnel house, walk-in troughs allow more sheep to be housed in a given tunnel.

Pole barns

Low-cost buildings do not need to be unconventional like 'topless' yards and plastic tunnels. A home-built pole barn is probably the cheapest conventional building. The design of pole barns is relatively simple and they have been widely used for many years for livestock housing, fodder storage and machinery sheds. If farm labour is available together with the basic carpentry skills, building a pole barn is not too difficult,

Building costs depend largely on the materials used and the cost of labour. The big saving made by DIY compared with using a contractor should be in labour cost — building workers are paid considerably more than most farm workers. However, unless secondhand materials or home-grown timbers are used the overall cost of DIY can be very close to that of a contractor. Because of bulk buying the contractor tends to have lower material costs, and skilled building workers will probably do the job faster. Pole barns are very suitable for the use of secondhand materials such as telegraph poles and corrugated steel sheets; costs increase by about 50 per cent when new materials are used.

Plate 23. A low-cost pole barn and roofless inwintering yard in Kent

Timber

Everyone understands timber; it is an easy material to work with and most of us can cut and join it. Durability and resistance to rot are important considerations. In dry conditions the risk of decay is small. However, timber exposed to the weather, in contact with the ground or with dung in exposed yards, must be protected.

Columns of rectangular cross section are more likely to decay than round poles, because they have a greater surface area for the rots to attack. If a pole is to be sunk into the ground it must be treated first. Secondhand telegraph poles will probably still be protected by the original preservative, but should be soaked in preservative to the required depth before use. They could then last up to fifty years. Posts can also be protected with concrete or by collars of plastic piping. If piping is used it should extend from about 150 mm below ground level to above the highest level of manure. New imported softwood and new home-grown poles should be pressure-impregnated with preservative. Treatment adds about 15 per cent to the cost of the timber.

Secondhand timber is cheaper than new and can often be purchased in greater lengths of larger cross-sectional area. Joists 50 mm × 275 mm were commonly used in the construction of old houses when the quality of timber was much better than it is today. Old nails and dirt-impregnated surfaces are the biggest drawbacks to secondhand timber. Dirty surfaces make gluing difficult and nails take time and perseverance to remove. It is also worthwhile checking that the timber is of consistent thickness.

New home-grown softwood is often available locally from estate sawmills. Timbers of varying cross-sectional size and length can be obtained, but they should be pressure-treated with preservative. The main advantage of home-grown timber is that it is clean and can be sawn to the required size. It is cheaper than imported softwood and the time and effort used in collecting, sorting and preparing secondhand timber are saved. New home-grown softwood poles 6 m long by 200 mm minimum diameter, pressure-treated with preservative, are likely to cost £20–25 each, compared with telegraph poles at around £5–7 each. But secondhand telegraph and electricity poles are not always available when you want them. Some thought has to be given to acquiring adequate supplies early in the planning process. There can be big savings over the price of new poles and so it is worth the effort. Larch poles cut from farm woods can also be used but should be pressure-treated.

Planning a pole barn

If you believe that you have the skill, time, labour and tools to plan the building, lay the foundations accurately, construct and erect the frame, fix roofing and guttering then select a design and obtain the materials. Designs vary enormously, there must be almost as many as there are barns. Many of the barns standing on farms today will have designs based on guesswork and rules-of-thumb. Some of the most common pole barn designs are illustrated in Figure 5.5. All are suitable for use as sheep houses, but C is more suitable for silage feeding down a central feed passage. Detailed information on the design and construction of pole barns is available from the Timber Research and Development Association (TRADA) and from ADAS. The ADAS *Land and Water Service Technical Note* 'Pole buildings, design and construction' is essential for the DIY planner/builder. As a step-by-step guide it is invaluable and covers every aspect from tree felling through pole sizes to roof fixing, for a range of buildings suitable for different areas of the country. The suggested designs all conform to BS 5502. Using this ADAS information most farmers should be able to erect pole barns which will last twenty years or even longer.

Exposure to high winds and snow loading is an important consideration. The map in Figure 5.6 shows areas of England and Wales which are considered high, medium, and low loading areas. The figures apply only at altitudes of less than 200 metres above sea level. Minimum pole diameters for low, medium and high load areas are given for various spans of pitched and monopitch roof. If buildings are to be erected on certain soil types, such as peat or soft clay, extra foundations may be needed.

The barn must be set out accurately on the ground, proper foundations must be excavated and base levels should all be the same after the post holes have been dug and the bottoms of the holes filled with concrete.

Trusses made of poles or sawn timber rafters are best made up one at a time on site so that they can be lifted straight into place with a tractor fore-loader and ropes. Monopitch buildings are easier to construct but need bigger timbers for an equivalent floor area.

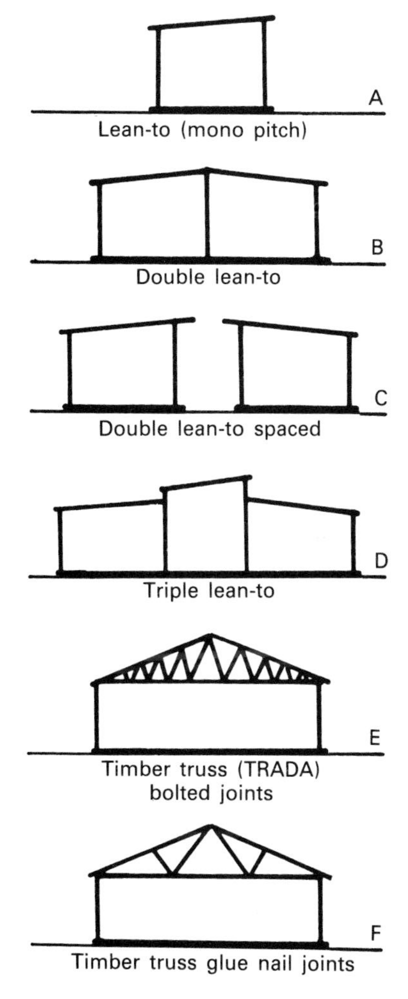

Figure 5.5. Possible designs of DIY pole barns
Source: East of Scotland College of Agriculture

The Rosemaund house

The commonest monopitch sheep house design is based on the low-cost house developed at Rosemaund EHF in Herefordshire in the 1960s. The design is simple, construction can be carried out easily by farm staff, ventilation is good, the sheep have a dry bed and mechanised silage feeding is possible. Each house is 7 m × 40 m and houses four groups of sixty ewes (1.1 m^2/ewe); this allows 150 mm trough space for ad lib forage feeding. Trough space is increased for concentrate feeding either by bringing in sheep troughs from the field or by using walk in troughs as pen divisions. Total trough space per ewe is therefore 450 mm.

This design, although it has proved effective over twenty years, has two drawbacks. There is less protection for the

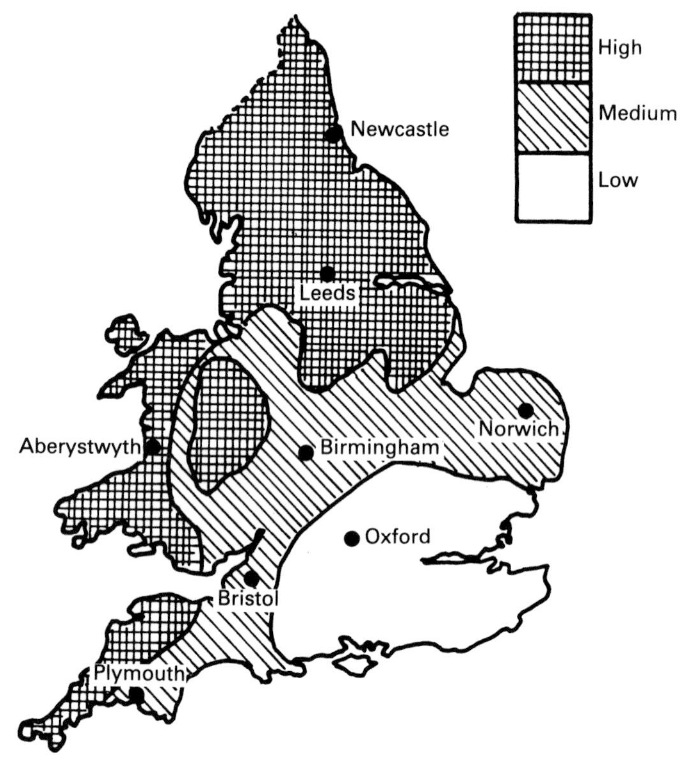

Figure 5.6. Map showing areas of high, medium and low loading
Source: MAFF, 1983, *Pole Buildings Design and Construction*

Plate 24. Open-fronted Rosemaund sheep houses sited for mutual shelter and silage feeding — Rosemaund EHF, Herefordshire *Farmers Guardian*

shepherd than in totally enclosed and roofed structures, and if the open front is exposed to wind and rain forage intake may be reduced because sheep shelter at the back of the building. Both problems can be overcome by careful siting or by placing two houses face to face for mutual protection.

Portal frame buildings

If you need a genuine, multi-purpose building with a high, clear span the portal frame structure gives this flexibility. Cost can vary from the cheapest DIY project using second-hand material, to prefabricated kits for self-erection or at the high-cost extreme a complete package deal with a concrete, steel or timber frame (Table 3.3).

Framed buildings have a number of advantages over traditional construction methods and the previously described specialist sheep houses. They can be prefabricated and mass produced in spans of up to 30 m; wide span and unlimited length give great flexibility of use. Because it has a large area of uninterrupted internal space, the building is very adaptable and therefore has a longer and more useful life than many other specialised structures. Erection of prefabricated kits on a prepared site is quick and relatively simple.

Most framed buildings are designed with a portal frame; the uprights and the rafters are designed as an integrated whole in either steel, concrete or timber.

Concrete frames are reinforced and pre-stressed, they are expensive but durable with a life of up to fifty years. Great strength, no maintenance, freedom from rots and pests, and resistance to mechanical damage are important characteristics.

Steel has similar advantages to concrete but does need to be painted at intervals.

Treated softwood can have a life equally as long as steel and concrete. Although not so resistant to damage, timber can be relatively easily repaired and lends itself to low-cost DIY construction. Information to enable DIY design and construction is available in the ADAS 'Pole buildings design and construction'. If you are considering the purchase of a build-

Plate 25. A steel portal frame house could have a useful life of forty years

Farmers Guardian

ing as a kit or as a complete package deal specify that the structure is designed to BS 5502.

BS 5502

Until 1980 designs for timber structures for farm buildings were based on BS 2053, which used only generalised loads for snow and wind in the design stage. BS 5502 however, requires snow loads and wind loads to be calculated for the exact location of the building. It follows therefore that a building at sea level in a sheltered position with low snow loading can be constructed more cheaply than a similar building on an exposed hill farm in the north of Scotland. Other factors relevant to the site such as soil conditions and class of building are taken into account. Farm buildings are divided into four classes (Table 5.4).

Table 5.4. Building classes in BS 5502

Class	*Design life in years for structural components*
1	50
2	20
3	10
4	2

Source: Taylor, N., 1983.

References

ADAS (1983), *Land and Water Services. Technical note. TN/FBS/22.*
Cornwell-Smith, M.S. (1981), *Farm Buildings Digest, 16,* 4, 5-6.
Dymond, A. (1978), *Farm Buildings Report*, FBIC.
Taylor, N. (1983), *Farm Building Progress, 73,* 1-2.
Watson, G.A.L. (1982), *Farm Building Progress, 70*, 7-21.

Chapter 6

FEEDING THE HOUSED EWE

ADEQUATE NUTRITION at mating, in late pregnancy and in early lactation is the key to successful management of intensive sheep enterprises. A greater opportunity to exercise control over nutrition is given if the ewes are housed rather than outwintered. Housed ewes should be fitter, their lambs should be stronger and should receive a better supply of milk and colostrum. Strong, well-fed lambs have a better chance of survival than weak, undernourished lambs so the lambing percentage should be higher. Although a greater proportion of lambs will survive, the number of lambs born will not be increased as a result of the greater degree of management control while the ewes are housed, because lamb numbers are largely determined by the ewe's body condition at mating and in early pregnancy. Feeding the ewe while it is housed cannot therefore be dealt with in isolation from the equally important period before housing.

NUTRITION BEFORE HOUSING

Ewes must be in good condition at housing to ensure that the maximum number of viable foetuses are produced. Improvement in condition is usually achieved by expanding the grazing area to 'flush' ewes before mating. The aim is to stimulate the ewe to produce a good supply of eggs at ovulation and to achieve the highest proportion of these eggs fertilised. The flock should be handled to assess condition two weeks before mating. The target condition score at mating is 3½. Any ewes in poor condition should be separated from the flock and given supplementary feeding, up to 500 g cereals daily. There is some evidence that the inclusion of a small amount of undegradable protein, such as fishmeal, in the ration at this stage helps to stimulate ovulation. In general the better the condition score, the better the ovulation rate. A high ovulation rate does not guarantee a high lambing percentage but it does increase the chances of a good lamb crop.

There is evidence that nutrition has an even greater influence on conception and foetal viability in ewe lambs. Ewe lambs should weigh at least 70 per cent of their mature body weight at mating. To meet the extra nutritional needs of ewe lambs, it is advisable to start supplementary feeding of a cereal/urea ration 0.5 kg daily, two to three weeks before mating. Avoid stress by providing ample trough space so that all lambs can feed at once.

Table 6.1. Effect of body condition score at mating on lambing percentage

	Body condition score at mating						
Breed	*1*	*1½*	*2*	*2½*	*3*	*3½*	*4*
Scottish Blackface	—	79	—	—	162	—	—
Welsh Mountain	60	65	105	116	123	—	—
Mule	—	—	149	166	178	194	192
Greyface	—	—	147	163	176	189	184
Scottish Halfbred	—	—	148	170	183	217	202

Source: MLC, 1982.

Hill breeds tend to show a much more dramatic response to improved body condition than the more prolific lowland breeds. Even when in poor condition Finn sheep have a high ovulation rate, but because of their inherently high prolificacy they must then draw on inadequate body reserves of fat in late pregnancy. Apart from its effect on ovulation, good condition at mating provides a store of body fat which can buffer the effects of some undernutrition in the later stages of pregnancy. A change in body condition of one unit represents a change in body weight of 10 kg which is a considerable store of energy.

In early pregnancy, body condition should be maintained for at least a month to guarantee the maximum survival of fertilised eggs. After fertilisation the egg floats in the uterine fluid for fifteen to forty days before it is implanted in the wall of the uterus. No abrupt changes in nutrition should occur at this time. Severe undernutrition, even for a short time, or gross overfeeding can reduce embryo survival.

Until the end of the third month there is very little foetal growth. At ninety days the foetus weighs only 15 per cent of the birthweight of a newborn lamb. The uterus, placenta and uterine fluids increase rapidly in weight so that their combined weight is from 4 to 6 kg at the end of the third month. A small loss in condition, up to half a unit of condition score, can be tolerated during months two and three without adverse effect. If some ewes are overfat (condition score 4), the loss of three-quarters of a unit of condition could lessen the risk of pregnancy toxaemia in late pregnancy. Ewe lambs still have a requirement for growth so body weight should increase by about 5 per cent between mating and the end of the third month of pregnancy.

Pre-housing action list

- Start to feed ewe lambs three weeks before mating.
- Condition score ewes two weeks before mating, target score is 3½.
- Separate ewes in poor condition, 2½ or less, feed a cereal supplement.
- After mating avoid abrupt changes in nutrition.
- Adjust stocking rate after mating to maintain condition.
- Do not allow ewes to lose more than half a unit condition score in months two and three.
- Condition score before housing.
- Separate ewes in poor condition, give them preferential treatment.

NUTRITION DURING HOUSING

The majority of flocks are housed for the last two months of pregnancy. Adequate nutrition at this stage is vital. The foetus makes 70 per cent of its growth in the last two weeks before birth. Nutrients are also needed to maintain the ewe, to support the growth of milk-secreting tissue and to ensure a good supply of colostrum. At the same time the efficiency of conversion of dietary metabolisable energy (ME) is only 13 per cent. Dietary energy requirements are therefore high. In fact, they are so high that it may not be possible to make up the deficit between dietary ME and the needs of the ewe.

Some limited and controlled mobilisation of body fat reserves can be used to make up the difference at this stage.

Table 6.2. Metabolisable energy requirements (MJ of ME)

		60 kg ewe	*70 kg ewe*	*80 kg ewe*
Mid-pregnancy (2nd & 3rd months)		7	8	9
Late pregnancy (5th month)	single	11	12.5	14
	twin	14	15.5	17
Lactation	single	18	19	20
	twin	22	23	24

A high protein concentrate containing protein of low rumen degradability, such as fishmeal or soya bean meal, reduces the effect of low energy intake on lamb birthweight. Colostrum production is also stimulated because the protein helps to mobilise and utilise body fat to make up the energy deficit in early lactation.

There are two sources of protein in a ewe's ration: non protein nitrogen (NPN) and dietary protein. The NPN breaks down to ammonia in the rumen and is then used as a protein source by the rumen micro-organisms. After it becomes microbial protein it passes into the lower gut where it is digested and absorbed. Dietary protein has a different fate. Some of it is broken down and turned into microbial protein but a fraction escapes. This is the undegradable dietary protein (UDP). UDP passes out of the rumen into the true stomach and intestines where it is broken down and absorbed directly as amino acids. The extent to which protein sources can be degraded in the rumen varies and has been measured. White fishmeal and heat-treated soya bean meal are sources of protein high in UDP and can be incorporated into home-mixed rations. Protein quality and quantity in many proprietary and farm mixed concentrates are often inadequate. Rations high in UDP, for use in the last six weeks of pregnancy and early lactation, are given in Table 6.3.

Underfeeding of energy is likely in most commercial systems despite concentrate feeding. To meet energy requirements as closely as possible, introduce a cereal supplement eight weeks before lambing. If high-quality hay or silage is

Table 6.3. Home-mixed componds high in UDP

% Fresh weight	*12% CP*	*14% CP*	*16% CP*	*20% CP*
Whole cereals	80	75	70	60
Dried molassed sugar beet pulp	10	10	10	10
Soya bean meal (45% CP)	7.5	12.5	17.5	27.5
Mineral, vitamin supplement*	2.5	2.5	2.5	2.5

* Include 12 g calcined magnesite per ewe two weeks before to four weeks after lambing.
Source: ADAS.

available (ME — 10.5 MJ/Kg/DM), supplementary feeding can be delayed until the fourth or fifth week before lambing. Concentrate feeding should be matched to the growth of the foetus by increasing the level of feeding progressively. The amounts of feed needed to meet the requirements of twin-bearing ewes are outlined in Table 6.4. These recommenda-

Table 6.4. Examples of amounts of feeds that supply the levels of energy and protein for ewes of two different liveweights carrying single, twin or triplet foetuses

Ewe weight:	*50 kg*			*70 kg*		
Foetuses:	*single*	*twin* (*kg per day*)	*triplets*	*single*	*twin* (*kg per day*)	*triplets*
Six weeks before lambing						
Hay*	0.83	0.83	0.83	1.00	1.00	1.00
or silage	2.6	2.6	2.6	3.5	3.5	3.5
plus concentrates	0.18	0.30	0.34	0.24	0.37	0.44
Four weeks before lambing						
Hay*	0.83	0.83	0.83	1.00	1.00	1.00
or silage	2.6	2.6	2.6	3.5	3.5	3.5
plus concentrates	0.28	0.45	0.51	0.36	0.56	0.66
Two weeks before lambing						
Hay*	0.83	0.83	0.83	1.00	1.00	1.00
or silage	2.6	2.6	2.6	3.5	3.5	3.5
plus concentrates	0.37	0.59	0.68	0.48	0.75	0.86

* Assuming ME concentration in the dry matter of 10.0 MJ for hay and 11.0 for silage, and dry matter content for silage of 25 per cent. Roots could be used to replace 75 and 50 per cent of the concentrates on a dry matter basis at six and two weeks before lambing, respectively.
Source: MLC, 'Feeding the Ewe'.

tions are based on good-quality hay or silage and cereals with 3-4 per cent fishmeal, to be fed until the last four weeks before lambing. During the last four weeks, protein levels should be increased by the addition of 8 per cent fishmeal or by using one of the high UDP mixes (Table 6.3).

If the ration contains a high proportion of ground barley or maize it should be given in two feeds a day, because some ewes will eat more than others and this can be dangerous. Finely ground cereal depresses the rumen pH and causes acidosis. Fibre digestion, roughage intake and appetite are all reduced which can result in pregnancy toxaemia. It is necessary to avoid extremes of diet and try to assess the effectiveness of feeding by condition scoring three to four weeks before lambing. The foetus is still only 60 per cent of its birthweight and there is a last chance to correct over or underfeeding. If the condition score of the ewes was 3 at ninety days after mating, a reduction of half to one unit of condition in the last eight weeks will not depress birthweights. However, it will have an effect on colostrum and milk production by depleting fat reserves. Therefore feed to maintain condition score at no less than 2½ four weeks before lambing and preferably 3½–4.

Ewe lambs and shearlings have a higher feed requirement as they are still growing, and so competition at the trough will be reduced if they are penned separately from the older ewes.

If all breeding sheep are colour-marked at mating, they can be penned according to their expected lambing date. Nutrient requirements can then be met more exactly. Overall feeding should be planned according to the expected lambing percentage. Unless an ultrasonic device for the detection of multiple foetuses has been used to identify single-bearing ewes, lowland ewes should be fed as if carrying two lambs.

Conserved forage

Sheep can be successfully fed on a wide range of fresh, ensiled or dried roughage. In practice, choice is limited by the farm system and what is available at a given time. Hay is still the commonest roughage for sheep but with increasing intensification, more and more farmers are turning to silage. Straw, regarded as a waste product on arable farms, can provide the basis of a highly successful winter ration, when balanced with an appropriate concentrate (Plate 27).

Hay

Too great a reliance on low-quality roughage leads to undernutrition, but ewes will winter well on even moderate-quality hay. They are able to select the leaf and leave the stem. The capital cost of haymaking and storage is low, transport and feeding are simple if laborious. However, bad weather at haymaking reduces the nutritional value and quality is highly variable. It is also difficult to combine haymaking with intensive management.

Table 6.5. Daily feeding levels in late pregnancy (70 kg ewe)

	Concentrates kg/day
Low-quality silage (55D, ME 8.0, 25% DM)	1.0
Medium-quality silage (60D, ME 10.0, 25% DM)	0.5
Very high-quality silage (65D, ME 10.5, 25% DM)	0.25

Silage

Silage-making is less weather-dependent than haymaking and quality tends to be higher and less variable than hay. Aftermath grazing is available earlier and silage is more compatible with intensive grazing. Once sheep are housed silage feeding becomes a practical proposition. Outdoor clamps of self-feed silage for outwintered sheep have been tried but rain, mud and the inevitable high wastage in small unwalled silos have not attracted many enthusiasts.

It is vitally important to provide high-quality silage. The sheep cannot be as selective with silage as with hay and material of 25 per cent dry matter and 10–10.5 MJ/ME/kg dry matter should be the aim. Feeding can be done with a forage box or silage block cutter, or big bale silage can be fed in bunkers in the house. Any unacceptable silage should be removed from the troughs each day but if the quality is high and fermentation is good, wastage will be minimal.

Plate 26. Silage blocks for sheep at Liscombe EHF

Plate 27. Oldenburg ewes feeding on wheat straw at Nackington, Kent

Straw

Results from ewes overwintered on untreated straw can equal those from ewes fed on silage or hay if an appropriate concentrate is fed to make up for the lack of energy and protein in the straw. Ewes should be condition score 3½–4 at housing, because some loss of condition is inevitable in the first week as they adjust to the ration. Do not feed any hay before straw feeding as this can lead to rejection of the straw. About 1.5 kg of clean, bright straw should be offered each day, decreasing to 1.0 kg nearer lambing. Ewes eat the straw selectively and there is some wastage; however, it should be available twenty-four hours a day so that they can eat little and often.

Straw provides less energy (6.0 MJ/ME/kg/DM) and less protein (8.0 g/dcp/kg/DM) than hay (8.0 MJ/ME/kg/DM; 45 g/dcp/kg/DM). Concentrates to balance the straw should contain at least 12.5 MJ/ME/kg/DM and 16 per cent crude protein (CP). A mix containing 80 per cent cereals, 20 per cent soya bean meal, minerals and vitamins should meet these requirements. Add a sheep mineral containing 15 per cent calcium and 10 per cent phosphorus to the concentrate mix at 25 kg per tonne. Calcium deficiency can be a problem with straw-based diets and the ratio of calcium to phosphorus may have to be changed by adding ground limestone nearer lambing.

Table 6.6. Concentrate allowances (kg) with ad lib straw

	Weeks before lambing			
Ewe weight	*up to 8*	*6*	*4*	*2*
60 kg	0.49	0.67	0.80	1.00
70 kg	0.54	0.75	0.87	1.14
80 kg	0.59	0.81	0.96	1.22

Source: ADAS, 1981.

Although more concentrates are needed with straw feeding, the total feed cost is lower. No hay needs to be made and this can release land for cereals or more sheep can be kept on the same area of grassland.

Feeding during housing action list

- Feed to maintain condition score 3½–4½ four weeks before lambing.
- Provide adequate dietary energy and protein.

FEEDING IN LATE PREGNANCY/EARLY LACTATION

The importance of body fat reserves and body condition scoring have been emphasised earlier in this chapter. Energy laid down cheaply as body fat in the autumn is stored during pregnancy to be drawn on during the last weeks of pregnancy; and even more important, in the early weeks of lactation. Adequate supplies of colostrum are essential for lamb survival and a high lambing percentage. Milk yield determines lamb growth rates, because during the first six weeks of life lambs depend entirely on the dam's milk supply. Only after six weeks does grass intake gradually replace milk. Ewe's milk is such a concentrated source of energy and protein that it takes 3.5–5.0 g grass dry matter to replace 1.0 g milk.

Milk yield depends largely on the ewe's intake of energy and protein, and the degree to which the protein is degraded in the rumen. The energy requirement of a ewe suckling twins is 70 per cent higher in the week after lambing than it was two weeks earlier. Underfeeding in early lactation leads to a lower peak yield, shorter lactation and lower lamb growth rates. After six weeks, underfeeding has less effect on yields. Ewes with a high potential milk yield tend to lose a great deal of weight in the first six weeks after lambing because their intake of grass and concentrates is insufficient for their needs. A supply of UDP is an important catalyst in the mobilisation of body reserves when dietary energy is in short supply. A ration high in fishmeal or soya bean meal is necessary until grass growth is adequate enough to meet the needs of milk production.

Work at the East of Scotland College of Agriculture has shown that when grass is 6–7 cms high, as measured with an MMB Grassmeter, supplementary feeding can be withdrawn. There is no advantage in feeding concentrates beyond this stage of grass growth.

Early lactation action list

- Continue to feed a concentrate ration with high UDP content.
- Withdraw concentrates when grass is 6–7 cm high.

Chapter 7

REARING LAMBS ARTIFICIALLY

THE NEWBORN lamb is only equipped to consume and digest liquid food. Although two-day-old lambs will pluck at herbage and suck a blade of grass they soon let it go again. Only after about two weeks of age do the lambs actually eat the grass. The choice of diet for the newborn lamb is therefore rather limited, it either has to suckle milk from the ewe or after having received colostrum it can be fed with artificial ewe milk replacer. Within half-an-hour of birth the lamb can normally stand, suckling begins and the lamb starts to gain weight two to three hours after birth. There are however, occasions and systems of sheep production where numbers of lambs must be reared artificially.

High lambing percentages achieved by better management and the use of prolific breeds usually means more triplet and quadruplet lambs. As far as possible triplets should be reared on the ewe or should be fostered on to a ewe with a single lamb. A successful technique for rearing triplets on their dam has been devised at Drayton Experimental Husbandry Farm (EHF).

Triplet rearing

The ewe and her three lambs are kept inside and closely supervised for at least a week. Ewe milk is supplemented with cold milk substitute from a bucket or bottle. After the first week the ewes and their lambs are turned out in small sheltered paddocks in small groups. Milk substitute is withdrawn as soon as the lambs are eating concentrates from a creep feeder. Triplets on this system grow more slowly (250 kg/day) than suckled twins (272 kg/day) and tend to reach slaughter weight and condition about two weeks later (Orskov, 1977). Only 1 kg of milk substitute and 7–10 kg of cereal/protein mixture per lamb is used on this system and it is considerably cheaper than artificially rearing one triplet (Table 7.4).

Artificial rearing

Normally, on most sheep farms artificial rearing is essential for at least some lambs, but there are sheep production systems where large-scale artificial rearing is regularly practised. In those countries where sheep milk is a highly saleable product or where it is used for cheese and yoghurt making ewe's milk is at a premium. Milk replacers are much cheaper than ewe's milk and artificial rearing is very attractive. Generally the cost of artificial rearing is directly related to the amount of milk substitute used.

Colostrum

It is essential that the lamb receives colostrum early in the first day of life before the gut becomes impermeable to the large protein molecules, which are essential for the provision of passive immunity from disease. The laxative effect of colostrum helps to expel the meconium plug in the gut and, of course, it is a highly concentrated supply of nutrients. Any lambs taken for artificial rearing should have received an adequate amount from the dam or 100 kg of colostrum should be administered with a stomach tube within two hours of birth (see Chapter 8). Where hygienic conditions are poor more colostrum will be needed and your vet may prescribe a broad-spectrum antibiotic for all artificially reared lambs, to give extra protection from *E. coli* etc. in the first week after birth.

Milk replacers

A great deal of research and development work has been done on the use of milk replacers for lambs in recent years. This has increased our understanding of the digestive processes in the newborn lamb and how the lamb can cope with different components in the milk replacer.

It was thought originally that milk replacers should closely resemble ewe's milk in their composition. However, recent findings indicate that the ratio of milk sugar (lactose) to the fat can be varied, and also that several non-milk fats, such as tallow and lard, are suitable for use in milk replacers if they are adequately homogenised. The non-ruminant lamb cannot digest starch and this should not be used to replace lactose; hydrolysed starch can, however, be added with little effect on lamb performance.

Table 7.1. The composition of ewe's milk and cow's milk

	%DM	
	Ewe's milk	*Cow's milk*
Fat	38	30
Protein	30	26
Lactose	25	37

Milk protein has a unique ability to clot in the abomasum which delays passage to the small intestine and thus the protein is more efficiently digested. If other animal or vegetable proteins are used to replace milk protein, this valuable quality is lost. Soya protein is poorly utilised if it exceeds 50 per cent of the protein in the replacer, but hydrolised, high-quality fishmeal protein can be used successfully as a complete replacement for milk protein. Most ewe milk replacers are formulated from skimmed milk powder with extra vegetable and animal fats.

Growth rate from birth to weaning is directly related to milk consumption, but where a creep feed is on offer this relationship declines. The high cost of milk replacer means that it is important to define the minimum level of milk replacer that will give a satisfactory growth rate. There seems to be little advantage in feeding more than 5 kg milk replacer. But, where milk intake is restricted and growth rate is low, there is a lower efficiency of conversion of solid food after weaning; this is most important as post-weaning diets are likely to be high in fibre. Research findings show there is no difference in growth rate when lambs are fed the same total daily milk allowance in two, three or six feeds.

Weaning

Milk replacers are usually far more expensive than cereal-based concentrate diets; this makes early weaning to solid foods highly desirable. Success depends on the state of rumen development at weaning. Until about three weeks of age lambs have little inclination to eat solid food. After three weeks, palatable solid food will be readily consumed but the amount eaten will depend on the amount of milk consumed. Solid food intake promotes rumen development, which in turn is under management control.

Milk intake can be easily restricted with artificial rearing and where creep feed is available lambs can be weaned to solid food at four to five weeks of age. As with artificial calf rearing, the rate of concentrate intake is the main determinant of the weaning time, but intake may be difficult to measure in practice. Liveweight is an unreliable guide to weaning, as heavier lambs may have poorly developed rumens as a result of greater milk intake. Age is the best guide; wean each group when the youngest lamb is a month old, and at least three times its birthweight.

Early-weaned lambs should have been introduced to a creep diet made up of highly palatable ingredients such as flaked maize, soya, rolled barley or wheat offals. The energy and protein content of the creep ration should be the same as that for calf creep feed. Beware of feeding calf creep pellets though, as the level of copper in the mineral fraction may be toxic to lambs. Proprietary early-weaning concentrates should contain 18 per cent protein, a high proportion of which should be rumen undegradable protein (see Chapter 6), which is required to maximise the performance of young, fast growing lambs.

Post-weaning nutrition

Lambs weaned at around a month old do not have fully developed rumens and this has an important bearing on the quality and nature of the post-weaning diet. Highly digestible diets which will pass quickly through the rumen overcome the disadvantage of small rumen volume. Cereal-based diets available ad lib allow the early-weaned lambs to perform very efficiently as converters of food into meat. When the diet contains roughages, these should be capable of rapid digestion, such as high-quality ground roughages and legume-based pasture. Fattening of lambs at this stage on poor-quality roughage is impossible.

Cereals do not contain enough protein to meet the requirements of early-weaned lambs. High growth rates and efficient food utilisation depend on the inclusion of protein supplements in the diet. As in the milk replacer a proportion of the protein should escape degradation in the rumen. Heat-treated proteins and proteins with low rumen degradability such as fishmeal, escape rumen degradation.

The ration should be based mainly on barley. Oats are too fibrous and where maize is used it should be restricted to a maximum 50 per cent of the grain portion. Work at the Rowett Research Institute has shown that a suitable protein supplement can be made up, as in Table 7.2, into 4 mm pellets for mixing with whole grain. When well mixed these pellets are consumed readily and there is minimal wastage.

Table 7.2. Protein, mineral and vitamin supplements for early-weaned lambs

10% inclusion *100 kg protein pellets* *+ 900 kg cereal*	*15% inclusion* *150 kg protein pellets* *+ 850 kg cereal*
80% fishmeal	85% soya bean meal
15% limestone	10% limestone
4% molasses	4% molasses
1% minerals/vitamins	1% mineral/vitamins

Trace minerals and vitamins per 1 kg total diet

150 mg	$ZnSo_4.7H_2O$
80 mg	$MnSO_4.4H_2O$
200 mg	MgO
5 mg	$CoSO_4.7H_2O$
1 mg	KLO_3
5,000 iu	Vit A
1,000 iu	Vit D
20 iu	Vit E

Source: East of Scotland College of Agriculture.

Feeding whole grain

Sheep feeding could be made simpler and less expensive if cereals were not rolled or processed before feeding. Oats, maize, barley and wheat can all be fed to sheep as whole grains more cheaply and with greater safety than when rolled or crushed. Ground cereals tend to increase the rate of fermentation in the rumen, increasing the proportion of propionic acid which leads to undesirable deposition of 'soft' subcutaneous fat. Whole grains are less likely to cause digestive disorders such as rumenitis. The sheep is very well equipped to grind grain slowly to its optimum size; let it do the job for itself.

Nutrition of older lambs

Late-weaned lambs and store lambs have a well-developed rumen and can cope with a more varied and fibrous diet. These lambs can, however, be much slower to adapt to intensive indoor finishing systems, so cereals should be introduced gradually. Protein supplementation is also needed for the efficient utilisation of cereal-based diets and low protein roughages such as hay, turnips, or where the fattening system is based on ad lib concentrate feeding indoors. High-quality protein of low rumen degradability is not necessary for lambs at this stage of development and cost can be reduced by the careful use of urea. The non-protein nitrogen in urea meets the needs of the rumen micro-organisms and stimulates high feed intake.

Urea must be thoroughly mixed into the cereal base of the ration. Where cereals are processed through a roller mill and mechanical mixer this may not be difficult. Even incorporation of urea into a whole grain diet requires a different approach. Liquid urea can be mixed with whole grains in an ordinary vertical mixer. The rates of inclusion of urea, minerals and vitamins are given in Table 7.3.

Table 7.3. Urea, mineral and vitamin supplements for addition to whole barley diets for store lambs

	Kg per tonne
Urea	10
Calcium chloride	12
Sodium chloride	3
Vitamin and trace mineral	1.25
Sodium sulphate	0.4

Source: East of Scotland College of Agriculture.

Preparation of the whole barley/urea diets

The calcium chloride, sodium chloride and urea should be dissolved in 30 litres of warm water (45°C). This solution is then slowly poured on to the whole grain in the mixer. Proprietary soluble mineral vitamin premixes can be obtained; 1.25 kg per tonne of grain should be mixed with 3.75 litres of water and added to the mixture. Finely powdered sheep minerals in suspension are also suitable. Finally, the fine crystals of sodium sulphate are added separately to the moistened grain. The concentrate ration will then have a crude protein content of 13 per cent and will only add £10–12 to the cost of the barley — a considerable saving on the cost of a purchased sheep concentrate.

Artificial rearing systems

The choice of a rearing system will depend on many factors including the number of lambs reared, season of lambing, cost of milk replacer and concentrates, and availability of labour. Rearing systems have been described in some detail in the MLC publication 'Artificial Rearing of Lambs', which

Plate 28. Lambs reared artificially on milk replacer and concentrate pellets
Volac, 1983

has been reprinted several times. The system should be simple and cost effective. Where possible, triplets in flocks lambing up to 180–200 per cent should either be fostered or reared on their mothers, as this is the most effective way of dealing with them. Some lambs born in litters to highly prolific breeds must obviously be reared artificially. Most shepherds usually end up with at least a few artificially reared lambs.

10 kg milk replacer plus concentrates (intensive)
Where small numbers of lambs are reared this is probably the simplest system to manage (Table 7.4). Lambs are weaned at four weeks after consuming 10 kg of milk replacer as ad lib cold milk, with a proprietary concentrate on offer from one week after birth. The cereal protein concentrate, or weaner concentrate, is then fed ad lib from three weeks until slaughter at approximately fourteen weeks. The system is convenient but expensive and may not have an adequate margin when lambs reach slaughter weight in late summer.

5 kg milk replacer plus concentrates
Although the amount and cost of milk replacer is reduced compared with the '10 kg system' the milk replacer must be fed warm and abrupt weaning at three weeks may severely affect performance. Weaner concentrate is on offer up to 15 kg liveweight, when the diet is changed to a cheaper barley-based ration.

Ad lib replacer plus barley/protein mix
On average each lamb consumes 35 kg of milk replacer on this system which makes it 50 per cent more expensive than the '10 kg system' and 100 per cent more expensive than the '5 kg system'.

10 kg milk replacer + concentrates + grazing
This appears to be the lowest cost system of all. The weaner concentrate is on offer from week one and lambs are weaned at four weeks. As early as two weeks of age the lambs can have access to young, leafy pasture, preferably with a high clover content and free of parasites. Total feed cost to slaugh-

ter depends largely on stocking rate and pasture management. The lambs should be rotationally grazed around four 1-week paddocks. Costings in Table 7.4 are based on a stocking rate of fifty lambs per hectare, at a lower stocking rate less creep feed will be necessary. The barley/protein pellet concentrate described earlier or high-quality dried lucerne pellets are a satisfactory creep feed.

Table 7.4. Artificial rearing systems

System	*Milk replacer* *kg/lamb*	*Weaner concentrate* *kg/lamb*	*Barley/protein mix 16% c.prot.* *kg/lamb*	*Cost* *£/lamb*	*Relative cost*
Drayton ewe plus triplets	1.0	10.0	—	7.74	100
10 kg milk replacer plus grazing*	10.0	5.5	23.00	12.67	164
5 kg milk replacer restricted	5.0	12.5	76.00	17.05	220
10 kg milk replacer intensive	10.0	75.0	—	21.48	277
ad lib milk replacer	38.0	—	5.00	33.00	426

Note: Milk replacer 0.85 pence/kg
Weaner concentrate 17.3 pence/kg
Barley/protein mix 14 pence/kg
* Assume stocking rate of 50 lambs/hectare.

Housing

Buildings for lamb rearing should provide good ventilation, they should be weatherproof for the benefit of the lambs and the rearer and be provided with infra-red lamps to keep newly-born lambs warm in cold weather. If the lambs are kept in relatively small groups of about eight or ten they are more easily matched for feeding and weaning, and they are also more easily observed and controlled.

Where lambs are reared outdoors on clean pasture in the spring and early summer, then simple straw bale shelters suffice once the lambs are readily drinking their daily milk replacer allowance.

Feeding equipment

A range of feeders is available from metal troughs to fully

automatic, self-cleaning, warm milk dispensers for up to 500 lambs. Although lambs will drink from a bowl or trough, it is difficult to keep lambs and dirt out of the milk. In practice most feeders rely on a teat, the simplest consist of a teat connected by tubes to buckets of milk replacer (Figure 7.1). A non-return valve in the tube makes it easier for the lambs to suck. Alternatively the teats are fitted near the base of a bucket or metal tank, these are easier to clean than tube and teat feeders but occasionally the teats are pulled out and the milk wasted.

Automatic, labour saving calf feeders have been adapted to feed large numbers of lambs and are capable of ad lib or restricted feeding of warm or cold milk. When the milk is fed cold, intake is slightly reduced but food conversion efficiency is not affected.

Sucking milk from any of the containers mentioned tends to be a group activity; too large a number of lambs per teat could result in competition for milk and restricted intake by weaker individuals. Under a 'restricted' feeding regime where milk is fed three or four times a day, one teat for each lamb is essential. Where milk is available 'ad lib' one teat to three lambs is adequate.

Artificial rearing checklist

- Rear triplets on their dams or foster one on to a ewe with a single.
- Choose the most appropriate rearing system and equipment to suit the number of lambs, season of birth and market.
- Ensure that the lamb has an adequate intake of colostrum soon after birth.
- Feed a proprietary 'ewe milk replacer' to weaning at approximately four weeks of age.
- Clean drinking water and 18 per cent crude protein weaner concentrate should be available from week 1.
- Wean when the youngest lamb in the group is four weeks old and weighs 3 × birthweight.
- Continue to feed highly digestible 'weaner' ration ad lib after weaning.

Further reading: MLC (1983), 'Artificial rearing of lambs'.

Chapter 8

THE FLOCK HEALTH PROGRAMME

HOUSING OFFERS opportunities for greater productivity and intensification of the sheep enterprise. But when sheep are kept in close confinement, the risk of a breakdown of flock health is greater and potentially more serious than with outwintered sheep. Although sheep may only be housed for three or four months of the year, if they are healthy outside, problems will be fewer when they come indoors. There are a number of excellent books devoted exclusively to sheep health. It is outside the scope of a 'sheep housing handbook' to catalogue all of the diseases to which sheep are prone. Instead, I have attempted to outline a twelve-month programme of preventive medicine. Successful managers anticipate problems, and nowhere is this more important than with housed sheep. Involve your vet at an early stage even before the house is built and draw up a flock health programme together.

Most vets only see sheep at lambing time, and often then only as a last resort. Vets tend to be treated as a 'fire brigade' service and are rarely involved in systematic preventive medicine for sheep. Sheep are no longer a low-cost, low-capital enterprise and this lack of veterinary involvement must be false economy. Regular advisory visits by your vet at critical times of year, such as pre-tupping, mid-pregnancy and pre-lambing should help to deal with problems before they arise. An essential preliminary is the production of a written health programme, preferably a chart or a monthly checklist appropriate to your own farm. Subsequent veterinary visits can then be used to check details of the plan and to review performance. The ADAS 'Shepherds' Calendar' is a useful guide.

Late summer — midwinter

The breeding cycle begins again after weaning and this is an appropriate time to start the shepherding year. This is the period when ewes are prepared for mating and the foundations of a profitable performance are laid.

Foot rot eradication

Foot rot is particularly important and should be dealt with now, before affected ewes are pulled down in condition and before the flock is housed. Foot rot can spread quickly through housed sheep if conditions under foot are warm and wet. Silage pulled into the pens from feeding racks results in a warm, wet floor where the sheep stand to eat. It is important to bear in mind the need for a dry floor when designing the house. The best treatment is eradication. Foot rot is a disease of the flock rather than individuals and therefore, treatment of lame, badly infected animals is merely a stop gap measure. Thoroughness, patience, perseverance and skill are needed to eradicate the disease. The method is simple but time-consuming.

Each sheep must be examined individually, so that it is probably best to start the eradication programme after weaning when numbers are lowest. If a start is made in early August, the flock should be free of foot rot by tupping and problems after housing will be minimal. Start on a dry day after a few wet days, because the feet will be softer and easier to pare. The routine is carried out on four successive weeks, preferably the same day each week, eg Friday.

Week No. 1

- Turn up each sheep and examine every foot.
- Pare infected feet, treat with appropriate aerosol spray.
- Mark infected sheep.
- Pass all sheep slowly through a footbath containing:
  10 per cent formalin (5 litres formalin to 45 litres water or 1 gal formalin to 9 gal water).
- Stand all sheep on clean dry straw over a hard floor for an hour.
- If possible return the flock to a field that has been free of

sheep for at least a week in dry weather, three weeks in wet weather.

Week No. 2

- Repeat the procedure outlined for Week 1.
- Mark infected sheep in a different place with a different colour.

Week No. 3

- Examine only sheep marked as infected at 1st and 2nd examination.
- Pass all sheep through the footbath as before.

Week No. 4

- Repeat the procedure for Week 3.
- Separate any infected sheep from the flock and give intensive treatment at intervals of a few days.
- Cull sheep which do not respond.

Week No. 10

- Examine the whole flock.
- If infection is found, even if only one foot, start again at Week 1.

Once the flock is free of foot rot, keep it that way. Isolate any bought-stock until they too are proved free of foot rot, disinfect your boots after visiting other sheep farms and consider vaccination. Vaccines do not cure foot rot but they do help to prevent infection. Eradicate the disease this year and start a vaccination programme next year.

External Parasites

Compulsory dipping against sheep scab in the autumn will also deal effectively with keds and lice. Until dipping against sheep scab became compulsory again after the reappearance of the disease in the early 1970s, there had been an increase in the incidence of other skin parasites. If not controlled, keds and lice can cause suffering and considerable economic loss after the ewes are housed.

Abortion

Perhaps it is a little early to think of abortion but we are planning ahead, and it may be advisable to vaccinate gimmers against enzootic abortion in the autumn. Infectious abortion causes severe economic loss every year, and the possibility of its introduction is the strongest argument for maintaining a closed flock.

Accurate diagnosis is essential, as early as possible in an outbreak. This can only be done by your veterinary surgeon and the Ministry of Agriculture Veterinary Laboratory after examination of the afterbirth, foetuses and blood samples from aborted ewes. All flocks experience sporadic abortion but if more than 2 per cent of ewes abort, you have a problem. Ten to 30 per cent of the aborted foetuses, and blood samples from ten to fifteen aborted ewes will be needed for laboratory examination. Some of the organisms can cause severe illness in humans: enzootic abortion, Q fever, salmonellosis and listeriosis are all potentially dangerous. It is an occupational hazard for shepherds and they should be aware of the dangers when handling samples for veterinary examination. The commonest forms of infectious abortion in the UK are enzootic abortion of ewes (EAE) and toxoplasmosis, followed by vibriosis and salmonellosis. Up to 50 or 60 per cent of ewes can abort depending on the type of infection.

Causes of abortion

INFECTIOUS	Virus	Border disease
	Chlamydia	Enzootic abortion of ewes (EAE)
	Rickettsiae	*Ehrlichia phagocytophila* (tick-borne fever) *Coxiella burnetii* (Q fever)
	Bacteria	*Campylobacter foetus* (vibriosis) *Salmonella spp.* (salmonellosis) *Listeria monocytogenes* *Corynebacterium pyogenes* *Brucella abortus*

	Fungi	*Aspergillus fumigatus* *Claviceps purpurea* (ergot) Various mycotoxins
	Protozoa	*Toxoplasma gondii* (toxoplasmosis)
NON-INFECTIOUS		Transport/handling stress Malnutrition Metabolic disorder — Pregnancy toxaemia

Source: Linklater, K. (1979), *In Practice 1, No. 1*, 30-3.

Enzootic abortion of ewes (EAE)

EAE or 'Kebbing' is essentially a disease of intensively managed low-ground ewes, and it is rarely seen under extensive management on the hills. A greater degree of intensification and sheep housing appear to have been associated with a marked increase in the incidence of enzootic abortion in recent years. At the same time new strains of the causative organism have developed, which may not be controlled or considerably reduced by the available vaccine. Losses can be as high as 30 per cent in the first year of infection, down to 15 per cent in subsequent years unless controlled by vaccination. Susceptible animals, usually lambs and gimmers must be vaccinated before they come into contact with infection and before their first mating.

Listeriosis

The bacteria responsible for this disease live in the soil. Poorly fermented silage contaminated with soil may lead to infection of the nervous system or abortion. Therefore only well fermented silage free of soil contamination should be fed to sheep.

Toxoplasmosis

The organism causing this disease (*Toxoplasma gondii*) is widespread, and can be introduced by a number of 'carriers'; man and cats have been implicated. If an infection occurs one year, it is unlikely to occur the next if flock replacements pick up the infection before mating. Infection in early pregnancy causes foetal death and abortion. Live lambs may be born after a late infection but the afterbirth is diseased.

Vibrionic abortion

Abortions caused by the organism *Campylobacter foetus* can occur in self-contained flocks and those which buy replacements. After an abortion storm, immunity is likely to last for several years. Crows, sparrows and carrier sheep have been implicated in the spread of this disease or it can also be introduced by wild fauna. It is very contagious and it is important to isolate infected ewes immediately. Once the flock is infected, very little can be done about it. After they abort ewes become immune and should be kept for further breeding. Outbreaks tend to be sporadic and localised so there has been no demand for a vaccine.

Abortion action list

If you suspect infectious abortion:

- Call in your vet immediately.
- Isolate aborted ewes.
- Remove and burn foetuses and afterbirth not needed for veterinary examination.
- Remember personal hygiene: material for veterinary examination is rich in infective organisms.

Selecting breeding stock

Any of the diseases mentioned in this chapter can reduce the breeding efficiency of the flock. It is just as important to select a disease-free animal as it is to select an animal of good 'type' for breeding. Animals without good teeth are usually rejected, but many with poor teeth are overlooked because only the incisors are examined. While the incisors are important for grazing, the molars are perhaps even more important. Without healthy molars the sheep cannot chew its cud properly and this leads to poor growth and lack of nutrients for the developing lambs.

In sheep with normal healthy teeth, the upper molars overlap the lower jaw. You can feel this overlap through the lips if you grasp the sheep's muzzle or run your hand along the jaw. The overlap should be smooth and curved, but if the molars are abnormal the curve is irregular. Even if the incisors appear to be perfect, the molars can be defective.

Just as most people examine the incisor teeth, most examine the ewes' udders for mastitis, teat size and placement, but few examine the rams thoroughly. Although raddle markers will identify infertile rams, it is then too late. The rams should be given a systematic veterinary inspection before mating. If their fertility is in doubt, semen can be collected and tested by your vet.

Late summer — autumn action list

- Eliminate foot rot and begin course of footrot vaccine.
- Compulsory scab dipping (also eradicates lice).
- Vaccinate gimmers against EAE, clostridial diseases.
- Check teeth of breeding stock.
- Inspect rams for fertility.
- Dose ewes and rams with anthelmintic before mating.
- Dose all sheep before they graze next year's pasture in the autumn.

Midwinter

The ewes will be housed during this period. Remember that sheep do not like wet feet and a damp atmosphere. Footrot should have been dealt with as described in the previous section. Pneumonia is the next major disease problem. Midwinter should be the time to plan a clean grazing system, not only for next year but for the next two or three years.

Winter shearing

If you have a prolific lowland flock and expect at least two lambs per ewe, it is worth considering winter shearing. This is a relatively new development in the UK and a considerable break with tradition. But there are some definite advantages if the flock is in good condition (at least condition score 2½) and lambing after mid-March in the south or early April in the north of England.

As the foetus develops the ewe's metabolism speeds up and more heat is generated at a time when the weather is becoming warmer. This heat is difficult to dissipate through the fleece, body temperature rises and respiration rate increases.

Plate 29. Romney Marsh ewes shorn two weeks after housing at Wye College, Kent

Winter shearing reduces this heat stress. Food is utilised more efficiently and lambs tend to be 0.5 kg heavier at birth. Heavier lambs have a better chance of surviving if they are triplets or twins. Up to 10 per cent more shorn ewes can be kept in the same area and trough space per ewe can be reduced 50–100 mm depending on ewe size.

On the other hand there are some drawbacks. The ewes will eat 10–20 per cent more, and very high-quality fodder may have to be restricted or there could be lambing difficulties with big singles. Treating vaginal prolapse is more difficult as there is no wool to tie the truss to and thus, a harness has to be devised. No great difficulty has been experienced with shearing at this time, but income from wool may be reduced the first year.

Pneumonia

Pneumonia is probably the greatest cause of economic loss to the UK sheep industry. A survey carried out by MAFF in 1964 showed that 82 per cent of nearly 10,000 sheep examined post mortem by Veterinary Investigation Centres had pneumonia. Almost always it is caused by the bacteria *Pasteurella haemolytica*. Stress or a primary virus infection usually triggers Pasteurella pneumonia. Outbreaks can be very severe. Where ventilation is poor and sheep are in close proximity, conditions are excellent for the spread of the disease. This feature should have been taken care of at the design stage. Housing should be dry, with as much air as possible above the sheep, draught-free at ground level and there should be adequate floor space. If relative humidity is high and the sheep have wet fleeces, improve the ventilation.

Sudden death or acute illness accompanied by shallow breathing, high temperature and nasal discharge are the usual signs of a pneumonia outbreak. Animals with milder symptoms often recover but never thrive.

Stress predisposes sheep to pneumonia. The disease can be caused by markets, travel in lorries, sudden change of diet or even a change in the weather. Introduce new diets gradually and avoid dusty or mouldy hay and bedding. Shearing at the housing stage can reduce the risk of pneumonia. Skin temperature and respiration rate are lower and this can reduce stress.

Pneumonia vaccines are available but their value has never been established. They cover only a limited range of the many serotypes of *Pasteurella haemolytica*. Think about the possibility of vaccination and take your vet's advice.

Maedi/Visna

Along with the importation of European breeds of sheep we have imported a form of virus pneumonia new to the UK. Maedi/visna is endemic in Scandinavia and many countries in western Europe. The disease has been responsible for the premature death of up to 20 per cent of ewes in some Dutch flocks. The virus has a long incubation period and is not usually seen until the sheep are three to four years old. There is no vaccine or treatment available and infected animals do not recover. Maedi spreads most readily when sheep are housed.

The first signs of the disease are rapid, shallow breathing after exercise. Then breathing becomes progressively more difficult until the animal dies three to eight months later. The disease eventually leads to paralysis of the hind-quarters in sheep over two years old.

Maedi/visna is brought into a flock by the introduction of apparently healthy but infected animals. Maintenance of a completely closed flock is not usually practicable, but infection can be avoided by purchasing replacements from a blood-tested flock. Where the disease is present it can be controlled; there are two methods. Either by regular six-monthly blood testing and culling of reactors, or by separating lambs from ewes at birth. The virus can pass from the ewe to the lamb via the milk. Lambs are then reared in isolation and regularly tested.

Orf

Orf is highly contagious and can occur at any time of year. Housed sheep may be especially at risk as the virus can live up to three years on the wood of pen divisions, hurdles and feed troughs. It is first noticed as scabs on the lips, often spreading to the nose and face. Once an outbreak has occurred, it can persist in a flock for a long time.

Secondary bacterial infections can occur particularly in lambs, which can pass infection to the teats and udder of the ewe. The bacterial infection can be controlled with antibiotic sprays but these have no effect on the virus. Vaccination may help to control the spread of the disease and aid healing in affected sheep. Vaccination with an attenuated living virus gives protection for up to six months. The vaccine should not be used unless there is good reason to expect an outbreak. If done in midwinter, it should be completed at least eight weeks before lambing to give the antibodies against the virus time to establish themselves.

Swayback

Swayback in lambs is caused by copper deficiency and is usually associated with hill and upland flocks. Incidence of the disease depends largely on weather conditions prior to lambing. In a mild winter with little snow cover between November and February, the ewes can graze and at the same time ingest soil with the grazed herbage. Soil inhibits copper metabolism and leads to swayback. In a severe winter when ewes are hand fed, less soil is ingested and swayback incidence is low. If hill ewes are to be housed after a mild period with little snow cover, proprietary copper injections at housing can be a useful insurance against swayback.

Clostridial diseases

The flock should be vaccinated against the common clostridial diseases: pulpy kidney, lamb dysentery, tetanus, Blackleg, Black disease, enterotoxaemia. If the vaccines are used correctly, they are completely effective. Breeding sheep should have two doses of a compound (7 in 1) vaccine in their first year, six weeks before lambing and a booster immediately before they lamb.

Planning the 'clean' grazing programme

Midwinter is the time to be thinking about worms in this year's lamb crop and planning for worm control. All grazing sheep are exposed to parasitic worms. Like the Unicorn, the 'worm-free sheep' is a mythical beast. Small numbers of worms are tolerated without much effect, but above a certain level growth is retarded. A realistic aim is to keep the worm burden small. There are up twenty species of roundworm which infect the stomach and intestinal tract of sheep. For practical purposes though, these will be treated as two problems: ostertagia and nematodirus.

Sheep housing is usually associated with high stocking rates and high stocking rates are only possible where parasitism is kept to an acceptable level. Adult sheep slowly develop immunity, it is the lambs and their productivity that are most at risk. As we are concerned with the success of the whole sheep system, during the housing period and in the grazing season, worm control is essential all through the year.

By combining a knowledge of the worms' life cycle with pasture management, and the strategic use of anthelmintic drugs, it should be possible to devise an effective control routine for any farm. Dosing alone will not control worms. Anthelmintic treatment of young lambs which are continuously exposed to infection has little effect on the worm burden. Worms that are removed are soon replaced.

Roundworms do not multiply inside the sheep. Eggs are deposited on pasture in the faeces of infected animals, the larvae hatch and are eaten by other sheep. These immature larvae burrow into the wall of the gut. At this stage they are more difficult to kill than mature worms. They may even be prevented from developing further by a build-up of immunity in the gut. Any radical change in the sheep's environment may however stimulate these 'inhibited' larvae to develop and emerge. Housing can provide the stimulus, as can a nutritional check. Adult worms in the gut produce eggs, peak production is in the spring and summer, this leads to an accumulation of eggs on the pasture. Larvae from these eggs emerge just as the lambs start to graze.

A four-year trial at Rosemaund EHF has found that ewes wormed with levamisole (Nilverm) ten days after housing produced 9 per cent more live lambs than untreated ewes in the same flock (Table 8.1).

Table 8.1. The effect of levamisole treatment on lambing percentage

	Lambing percentage	
Year	*Untreated*	*Treated*
1978	181	194
1979	188	192
1980	184	194
1981	178	187
Average	183	192

Source: Brown, D.C. *et al* (1984).

During pregnancy, even low levels of roundworm infestation can depress appetite and nutrient intake, and poor nutrition during the last three months of pregnancy causes increased foetal death and resorption.

Worm egg output from ewes rises rapidly around lambing, and anthelmintic treatment has been recommended to coincide with this explosive increase in worm eggs to avoid pasture contamination. However, recent work in the USA has shown that anthelmintic treatment six weeks before lambing under conditions where reinfection is unlikely (ie housing) eliminates the rise in worm egg numbers at lambing.

Treatment with levamisole soon after housing appears to have a double advantage over treatment at lambing. The potential for a higher lambing percentage is increased and the egg output is reduced.

Nematodirus

The eggs of nematodirus develop slowly and remain on

Plate 30. Ewes should be wormed ten days after housing *Farmers Guardian*

pasture throughout the winter. Hatching depends on a sequence of cold followed by warmer conditions. The size and timing of the hatch therefore varies from year to year. In some years as the temperature rises in May and June there can be an explosion of hatching larvae. Lambs up to four months old are most susceptible, the effects being sudden and fatal.

Ewes excrete relatively small numbers of nematodirus eggs. Some of these hatch in summer and infect the lambs. Worm numbers are low but their egg output is high. Eggs accumulate through the summer and autumn, storing up trouble for next year and so, infection is transmitted from one crop of lambs to the next.

The most effective way to tackle the problem is to prevent the ewe from depositing eggs on 'clean' pasture, and then to graze lambs on 'clean' pasture. On a new ley, first-year grass that has not previously been grazed by sheep, there is no danger to lambs from nematodirus. If the pasture has only been grazed by cattle or used solely for conservation in the previous year, it too can be considered 'clean'.

An effective routine is as follows:

- Dose all ewes ten days after housing or immediately before they go to 'clean' pasture.
- Dose lambs on 'clean' pasture at the end of July to eliminate the small number of worms present.
- Dose all other sheep using the pasture in autumn and winter before they are moved on to it.

Where lambs must be turned out on to older leys or grassland where the above routine has not been followed, use the same procedure but dose the lambs at the end of May, in late June and at the end of July.

Ostertagia

Overwintering eggs are not quite so important as with nematodirus but they cannot be ignored. The overwintered eggs can cause worm burdens in lambs in June and July that are not effectively controlled by dosing alone. A twelve-month, year-round, management plan is needed. Any ewes that are going to graze next year's sheep ground in the autumn should be dosed beforehand. If possible, keep these fields clear of ewes and lambs until after mid-June when most of the larvae will have hatched and perished.

Through a combination of pasture management and routine dosing, pastures can be kept relatively clean. Treatment will be more effective and the cost of drugs reduced. If the dosing routine suggested is used in conjunction with 'clean' pastures it will give control of nematodirus and ostertagia.

EWES

- Dose rams six weeks before mating.
- Dose ewes before mating.
- Dose rams after mating.
- Dose ewes two weeks after housing.
- Dose before moving anytime to 'clean' grass, eg hay or silage aftermaths.
- Dose any sheep before they graze next year's ewe and lamb pastures in the autumn. Do not graze these fields until mid-June, the following year.

LAMBS

- Dose in July on clean pasture, in mid-May, late June and end of July on other pastures.
- Dose before moving to 'clean' pastures.
- Dose at weaning and move to 'clean' pasture.

Any bought-in sheep should be dosed on arrival or before moving on to 'clean' pasture.

The provision of clean grazing should really be planned two or three years ahead. Clean grazing can be achieved by rotating blocks of grass between cattle, conservation and sheep. Using undersown, three-year leys, the stubble can be grazed by dosed ewes and then used for ewes and lambs in the spring. After weaning the ewes continue to graze the first-year grass and lambs are moved to aftermaths on second-year grass. Third-year leys can be grazed by ewes and lambs.

Lambs born before mid-March are at risk from overwintered ostertagia, but lambs born after this date are more at risk from nematodirus. Ostertagia becomes a problem after July. Sheep grazing can be alternated with cattle grazing to provide clean grazing for both cattle and sheep. On three-year leys, year one cattle grazing followed by year two sheep and then year three conservation with calves and lambs on the aftermath, will provide clean grazing for both classes of stock.

If all pasture cannot be conserved, or on farms where there is permanent pasture, the grazing can be divided into two blocks and alternated between sheep and cattle. Conservation and grazing by weaned lambs is then confined to the more accessible fields.

Winter action list

- Consider winter shearing.
- Dose ewes with anthelmintic two weeks after housing.
- Vaccinate against pneumonia, clostridial disease, orf.
- Inject ewes with copper preparation if swayback risk is high.
- Walk ewes through 5 per cent formalin footbath every 10–14 days.
- Plan next year's clean grazing programme.

Late winter — spring

A recent survey of sheep deaths in Scotland showed that over half of all ewe deaths occurred around lambing time. Between lambing and weaning, lamb mortality is about 15 per cent with a range from 5–20 per cent between farms. Clearly many of these deaths are avoidable by better management. Careful feeding, especially in late pregnancy, seems to be the key factor in preventing ewe and lamb deaths. Underfed ewes give birth to undersized, weak lambs, the onset of lactation is delayed and milk yield is reduced. When feeding is adequate lamb mortality is more affected by litter size and the age of the ewe, but it is still largely under management control. Aim to keep the ewes in good condition at this time of year; condition score 3½–4 at lambing to ensure that lambs weigh at least 2.5–3.0 kg at birth, and that they have early and adequate intake of colostrum.

Three-quarters of all lamb deaths occur in the first four days after birth. Even when ewes are housed at lambing,

Plate 31. Twin lambs should weigh between 2.5 and 3.0 kg at birth for maximum survival

starvation/exposure can be a killer if there is not enough milk. Despite good hygiene, high levels of bacteria can build up and *Escherichia coli* will present a severe challenge to the weak lamb. Most lambs are unaffected by this bacterial challenge because they are strong and well fed.

When dead lambs were examined post mortem at Drayton EHF, 72 per cent had received little or no colostrum. Mortality in lambs weighing less that 2.7 kg was 50 per cent compared with 7 per cent mortality in lambs over 3.6 kg. It therefore follows that extra attention should be given to triplets. With prolific breeds, where the aim is to produce at least two lambs per ewe, there will be a large number of multiple births. There is good evidence that winter shearing of ewes reduces lamb mortality. On average lambs from shorn ewes are 0.5 kg heavier than lambs from unshorn ewes. At Drayton twin losses were halved and triplet losses cut by a third after winter shearing.

Table 8.2. The causes of lamb mortality

	Percentage deaths		
Cause of death	(1) *North of Scotland*	(2) *S.E. England*	(3) *Drayton EHF*
Stillbirth	18	35	25
Starvation/exposure	34	30	29
Bacterial infection	11	14	17
Other	37	21	29
	100	100	100

Source: (1) JOHNSON & MACLACHLAN, (1980), *Veterinary Record 106*, 238-40.
(2) PURVIS *et al*, (1979), *Veterinary Record 104*, 241-2.
(3) MAUND, (1977), *Drayton EHF Ann. Rev.*

Preventing ewe deaths

If the vaccination programme has been satisfactory and internal parasites are under control, the commonest causes of ewe death are nutritional. When ewes are housed, careful feeding in late pregnancy should largely prevent twin-lamb disease.

Pregnancy toxaemia (twin-lamb disease)

The incidence of twin-lamb disease in outwintered ewes is closely related to weather conditions. Low winter temperatures inhibit grass growth leading to undernutrition. High rainfall and low temperature provide climatic stress and twin-lamb disease is then more common. It is energy deficiency in twin-bearing ewes that causes the disease. Stress, such as change of diet, lack of trough space and overcrowding often triggers the problem. Regular condition scoring helps to identify unthrifty ewes which can then be penned separately, and given extra feed. Overfatness must be avoided too; aim for a condition of 3 in mid-pregnancy, rising to 3½–4 at lambing. If ewes are penned according to stage of pregnancy, rationing can be more precise and economical. It is vitally important to have adequate trough space, and feeding concentrates before roughage also helps. The roughage should be of good quality especially during the last four weeks of pregnancy. These measures to prevent twin-lamb disease also help to prevent vaginal prolapse.

Hygiene

Good hygiene at lambing is essential. Dirty hands and lambing ropes inserted into a ewe can cause metritis (inflammation of the reproductive tract), or even gangrene. Wash hands, soak ropes in disinfectant and insert antiseptic pessaries after an assisted lambing.

Hypomagnesaemia (grass staggers)

When ewes and lambs are turned out to heavily fertilised pastures, grass staggers is a major cause of death. Include calcined magnesite in the concentrates so that each ewe receives 14 grammes each day or dust the grass with calcined magnesite. Avoid applying potassium fertilisers in the spring as this locks up the magnesium in the soil and makes it unavailable to the sheep.

Preventing lamb deaths

High mortality among newborn lambs is not inevitable, even with prolific ewes. The main causes of lamb death are shown in Table 8.2; they are stillbirth, starvation/exposure and bacterial infection. Strong, healthy, well-fed lambs are not usually affected by exposure and bacterial infection. Colostrum provides concentrated energy to maintain body temperature and combat hypothermia (exposure), and the antibodies it contains protect the lamb from bacterial infection. Not all stillbirths are preventable, but close observation and skilled assistance at lambing reduces the incidence of death from dystokia. Comprehensive instruction in lambing skills is available through the Agricultural Training Board.

Preparation for lambing

Start to prepare for lambing well in advance; have everything ready at least two weeks before the first lamb is due. Although the normal gestation period is twenty-one weeks, some ewes will lamb a week early, so be prepared. Have extra labour available from other farm departments or by employing part-time staff.

Mismothering is the biggest potential problem with indoor

Plate 32. At least one individual pen may be needed for every six ewes at lambing *Farmers Guardian*

lambing. Round-the-clock supervision is time and money well spent. A heated hut or caravan on site with water, heat, light etc., should be provided. Do not economise on lambing pens. At least one pen to six ewes will be needed, more at the peak of lambing. These should be big enough to accommodate the ewe and lambs and to allow her to be assisted at

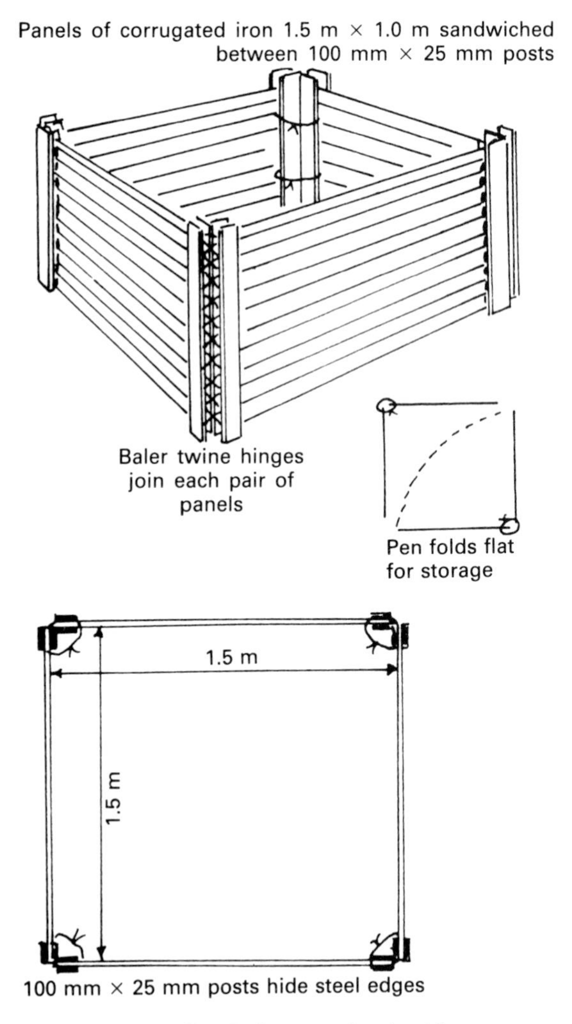

Figure 8.1. Simple free-standing lambing pen
Source: Harper, A.D., 1982

lambing, 1½–2 square metres should be enough. Clean buckets for water and feed are needed for each pen. Ewes and lambs should be penned together to 'mother up' for 24–48 hours after lambing. Castration and tailing with rubber rings can be done in the same pen during the first thirty-six hours. Remove afterbirths, and clean, disinfect and bed pens between ewes. A concrete floor and a convenient stand pipe are a great help when cleaning pens.

Colostrum

Colostrum is vital for survival. Lambs should suckle the ewe as soon as possible, at least within six hours of birth. Weak lambs can be given colostrum through a stomach tube about 100 cc per feed. Ewe colostrum contains immunoglobulins which give the lambs resistance to disease, and energy in the fat and carbohydrate is for heat production. Colostrum can be taken from early lambing ewes and stored in a freezer for use in an emergency. Yoghurt cartons make useful freezer packs of colostrum. Cow colostrum is almost as good, but avoid using milk replacer during the first forty-eight hours as large amounts could cause dehydration and death. Warm the frozen colostrum slowly to blood heat before use as too rapid heating can cause coagulation; a double saucepan does the job satisfactorily.

Navel dressing

At birth navels should be treated with veterinary iodine to prevent abscesses and joint ill. The iodine is best transferred from glass bottles to an old washing up liquid bottle. Leave one iodine container in each section of the house for immediate use.

Fostering

Every ewe should be turned out with two healthy vigorous lambs. When triplets have all suckled, the strongest lamb should be removed and fostered on to a ewe with a single lamb. Proprietary fostering pens are available, however home-made ones are just as effective (Plate 34). Plan to have one fostering pen for every thirty ewes. The ewe is restrained by the neck for two days after the lamb is introduced. When

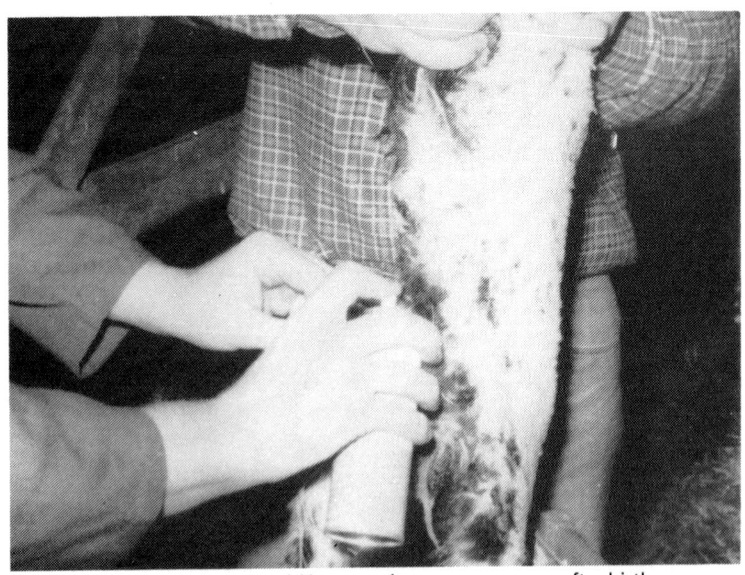

Plate 33. Navel dressing should be a routine measure soon after birth
Farmers Guardian

Plate 34. A lamb fostering pen made from pallets *Farmers Guardian*

the ewe has accepted the lamb they can both be placed in an ordinary lambing pen. Ewes with foster lambs can then be penned together for a few days before being turned out in small groups. Check them regularly for mismothering.

Ewe lambs should be very closely supervised. They take longer to lamb than ewes and their lambs can be slow to suckle. Inexperience, prolonged lambing and excitement may cause a ewe lamb to reject her offspring. If possible, pen before lambing and supervise until the lamb has suckled. Once 'mothered up' ewe lambs tend to be more maternal than shearlings, but it is wise to graze small groups of ewe lambs separately from the ewe flock.

When ewes and lambs are 'mothered up' and ready to turn out, number each family group using a different coloured aerosol for singles, twins and triplets. Dose the ewes for worms in the lambing pen. Do not transport too many ewes and lambs at once, keep family groups in separate compartments and unload each family in a different part of the field. A sheltered 2 or 3 acre paddock is needed for the first few days. This helps prevent mismothering and eases the shepherd's job.

Starvation-exposure

After stillbirth, starvation/exposure is the biggest killer of lambs, accounting for a third of all lamb deaths in the first two weeks of life. The normal body temperature of a lamb is 38-39°C, and hypothermia occurs when body temperature drops below this level. Hypothermia cases fall into two categories:

- Newborn lambs, up to five hours old, can suffer from severe heat loss caused by rain, wind and cold.
- Older lambs between five and seventy-two hours old cannot generate enough heat to maintain body temperature if they are starved, and have low blood glucose levels.

A highly successful resuscitation technique has been developed for both of these categories of lamb. Early diagnosis is essential for resuscitation to be effective and cannot be done without a thermometer. An ordinary clinical thermometer can be used or a more specialised electronic device can

Plate 35. The use of an electronic thermometer to diagnose hypothermia

Farmers Guardian

Plate 36. After drying with a towel, lambs with moderate hypothermia are given colostrum with a stomach tube *Farmers Guardian*

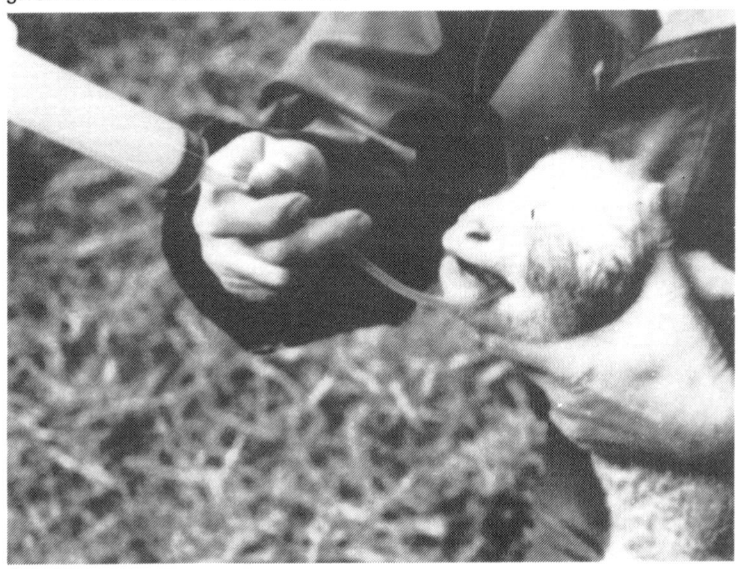

be purchased. Treatment depends on the age and temperature of the lamb.

Lambs of any age suffering from moderate hypothermia (37-39°C) should first be dried with a towel and then given colostrum with a stomach tube. The lamb is then returned to the ewe and both should be placed in a sheltered pen.

When severe hypothermia occurs (temperature less than 37°C) in newborn lambs less than five hours old, they should be dried and then placed in a warming box. Older lambs with temperatures less than 37°C usually have low blood glucose levels: an injection of glucose solution is necessary before warming. Ten ml of 20 per cent sterile glucose solution is most effective. The glucose is administered with a 50 ml syringe fitted with a 19 g needle. Insert the needle into the body cavity 1 inch to the side and 1 inch behind the navel at 45 degrees to the skin, pointing at the lamb's rump. Lambs with severe hypothermia are then re-warmed in air at 40°C after drying with a towel. Warming a wet lamb can cause further severe heat loss by evaporation.

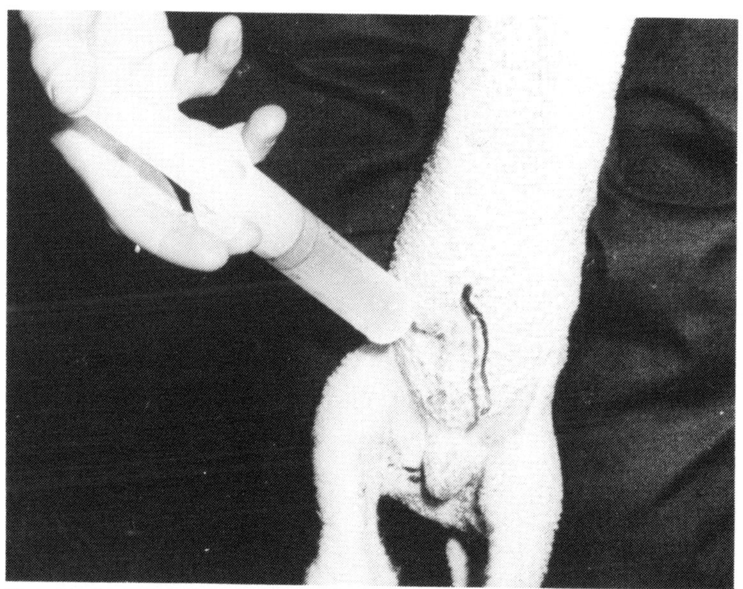

Plate 37. A glucose injection should be given half an inch to the side and one inch behind the navel

A cheap and effective warming box can be made with straw bales, weldmesh and a 3 kw domestic fan heater. Make the base with four bales and insert a steel channel on one side for the fan heater. Half-inch weldmesh is laid on these bales to make a floor and then the pen is completed with another four bales around the edge of the weldmesh. Place a sheet of polythene over the top to retain heat. Temperature can be regulated by the settings on the fan heater or by changing its position. Check the lamb's temperature regularly and remove from the box when body temperature reaches 38°C.

After removal from the box, intensive care should continue for lambs that are still weak. Cardboard boxes about 2 foot square and lined with newspapers make good disposable aftercare pens if an infra-red lamp is suspended about 1.3 m above the lambs. Give 100–200 ml colostrum by stomach tube after warming and feed this way three times a day until the lambs are sucking vigorously, when they are either returned to the ewe or reared artificially. An oral antibiotic adminis-

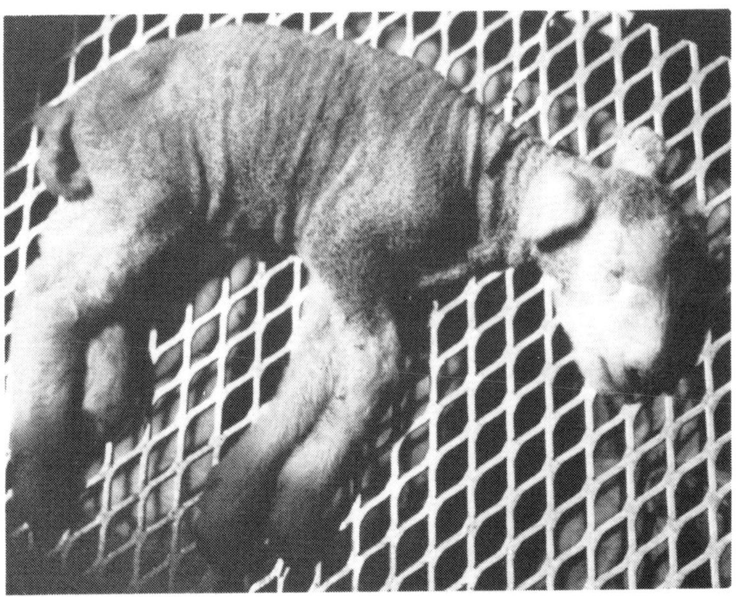

Plate 38. Lambs with severe hypothermia are placed in a warming box at 40°C after drying *Farmers Guardian*

tered twice daily is an insurance against infection in these lambs, but should only be done after taking veterinary advice.

Shepherd's equipment and materials for lambing

LAMBING EQUIPMENT
Paper towels + dispenser
Plastic buckets × 6
Disinfectant (Dettol)
Lubricant
Lambing ropes × 4

INTENSIVE CARE UNIT
Towels
Warming box
Stomach tube × 2
Frozen colostrum
Cardboard boxes
Infra-red lamps

MEDICINES
*Oral antibiotic
Antiseptic pessaries
*Antibiotic injections
Clostridial vaccine
Anthelmintic drench
Calcium borogluconate
20 per cent sterile glucose solution
Multi-vitamin injection

MISCELLANEOUS
Torch + batteries
Record sheets + notebook
Weigh scale
Ear tags + applicator
Marking stick & aerosols
Drenching gun
First-aid kit

SURGICAL EQUIPMENT
Sterile syringes: 2 ml, 10 ml, 20 ml, 50 ml
Needles: 18 g × 18 mm, 19 g × 25 mm, 16 g × 25 mm
Prolapse retainers
Rubber rings + applicator
Foot trimming knife
Dagging shears

* *On veterinary prescription*

Late winter — spring action list

- Ensure that feeding is adequate in late pregnancy.
- Condition score ewes six weeks and two weeks before lambing. Lowland ewes 3½–4, Hill ewes 3–3½ at lambing.
- Prepare pens and material at least two weks before the first lamb is due.
- Maintain strict hygiene when assisting lambing and in the lambing pens.

- Ensure that lambs receive adequate colostrum.
- Dress navels at birth with iodine.
- Pen ewe and lambs at least 36–48 hours before turnout.
- Castrate and tail lambs with rubber rings within 36 hours of birth.
- Dose ewes with anthelmintic before turnout unless wormed after housing.
- Give each family group a unique mark.
- Feed calcined magnesite to prevent staggers before and after lambing.
- Turn out to clean, sheltered paddocks in small groups.
- Have enough people to supervise lambing round the clock and ewes and lambs after turnout.

Summer

A clean grazing system was devised during the winter. The ewes were given an anthelmintic dose after housing or in the lambing pen before turn out. This combination of clean grazing and strategic use of anthelmintics will deal effectively with ostertagia and nematodirus. Vaccination against clostridial diseases if done correctly should protect the lambs until weaning. Disease problems should be minimal during the summer but ewes can still be lost with mastitis, and coccidiosis is becoming a problem with lambs born indoors.

Coccidiosis

Coccidiosis may be a problem three or four weeks after turnout. Any lambs suffering from severe scouring should be suspect. The incidence of coccidiosis has been on the increase in recent years and is thought to be associated with lambing indoors A sulphamezathine drench can be used but this is very laborious. Cattle feeding blocks containing monensin sodium have been successfully used to prevent and control the disease, but these blocks should only be used after consultation with your vet.

Mastitis

At weaning ewes should be dried off carefully to avoid post-weaning mastitis. Various estimates have been made of the number of ewes culled annually for mastitis, they range from as low as 5 per cent, to 16 per cent on some farms. The sheep house can be used again at weaning to dry the ewes off. Pen them inside for two days with only poor quality roughage and water. There is a double advantage in this method because the ewes are penned securely out of sight and sound of the lambs. Where the incidence of mastitis is high, there may be some benefit from treating the ewes with long-acting, dry-cow antibiotic tubes after weaning. Mastitis incidence has been reduced by this method but is expensive at £0.90–£1.00 per ewe.

Summer action list

- Watch for coccidiosis in lambs and treat accordingly.
- Winter shorn ewes can be dipped in late May.
- Dose ewes with anthelmintic at any time before moving to 'clean' pasture, eg. hay or silage aftermaths.
- Dose lambs on 'clean' pasture with anthelmintic in mid-July, and at weaning before moving to 'clean' pasture.
- Vaccinate lambs with multivalent clostridial vaccine at weaning.
- Dry off ewes carefully at weaning to avoid post-weaning mastitis.

References

Brown, D.C., Allen, M.L., and Ward, C.J. (1984), *Veterinary Record 114*, 58-9.

Eales, A. *et al* (1982), *Veterinary Record 110*, 118-20.

Eales, A. *et al* (1982), *Veterinary Record 110*, 121-3.

Eales, A. *et al* (1982), *Veterinary Record 111*, 451-3.

Hindson, J. (March 1982), 'Sheep health schemes', *'In Practice'*, 53-7.

Johnson, W.S. and Maclachlan, G.K. (1980), *Veterinary Record 106*, 238-40.

Maund, (1977), *Drayton EHF An. Rev.*

Purvis, *et al* (1979), *Veterinary Record 104*, 241-2.

Chapter 9

PLANNING A SHEEP HANDLING UNIT

IN ORDER to carry out a flock health programme successfully, thoroughly, and on time an efficient sheep handling unit is needed. The unit should be designed and sited to allow the flock to be held and treated in economically sized batches, with minimum stress on the sheep and minimum fatigue on the part of the shepherd. Inwintering requires regular footbathing and handling through the winter, therefore it is convenient for the handling pens to be an integral part of the sheep housing complex. On the other hand, during the grazing season it is most convenient to have handling pens and dipper central to the sheep grazing. Lightweight, prefabricated, demountable penning systems have an advantage here in that the hurdles and raceways can be moved near to the housing in winter and back to the grazing area in summer.

Siting

Siting is of utmost importance and all options should be examined before putting pen to paper to plan the system. Keep the unit as central as possible to the grazing area and to road access for loading sheep transporters. Because the flock may be standing in pens for quite long periods in the summer, shade from mature trees is an advantage, as is shelter from driving sleet and rain in winter. A good water supply is needed for rapid filling of dip baths and for cleaning down concrete areas. With such a high concentration of sheep, the resultant dung, urine, treading and drainage of water onto pen surfaces call for a free-draining base. It may be necessary to remove the topsoil and replace with hardcore or broken stone 70–150 mm deep, blinded with small aggregate or quarry waste. As sheep have a preference for moving uphill to escape, sheep flow will be improved if the pens are laid out on a slight incline, and this will also help the drainage of the site.

Sheep behaviour

Sheep have certain behavioural characteristics which can be exploited in the siting and design of the handling pens. Instinctively they move away from danger uphill, towards the horizon and back in the direction they came from. Try not to site the exit from a race near a building or wall, as there should be an illusion of freedom at the end of the passage. A slight curve in the race also aids the flow of the sheep. Sheep are highly social animals, they will walk, run, graze, bed down together and follow one another. In a natural flock the oldest ewe takes the lead, followed by her lambs then followed by her descendants of previous years. There is therefore a strong instinct to move towards another sheep.

Working conditions

Just as inwintering improves the shepherd's working conditions when the weather is at its worst, some thought should be given to the working area within the handling pen set up. The main working area can be part roofed at little expense. Design for minimum effort; gates and latches should operate smoothly and by remote control where necessary. Floors should be smooth, dry and clean but non-slip, a concrete slab for the dosing pens is essential.

Principles of handling pen design

Keep the pen layout as simple as possible, pens should be no larger than necessary and provision should be made to recycle the sheep back to the gathering pens. All surfaces that the sheep are likely to touch should be smooth and free of sharp edges or splinters, so keep the sheeting on the inside, posts on the outside. Although some concrete is needed in the handling area for the draining pens, it is unnecessary elsewhere — keep concreted areas to a minimum. There seem to be two basic pen layouts; pens with a shedding race at the side and pens with the shedder in the centre.

Plate 39. A well-designed handling unit, with minimal concrete use around the working area, handling and draining pens only *Farmers Guardian*

Plate 40. The exterior plywood panels provide an excellent splinter-free surface *Farmers Guardian*

Side shedding race (Figure 9.1)
The gathering pen has access to the shedding race in one corner, with a shedding gate opposite each side pen. At the opposite side sheep can be passed from the gathering pen into the drawing pen and on into the catching pen for the dipper.

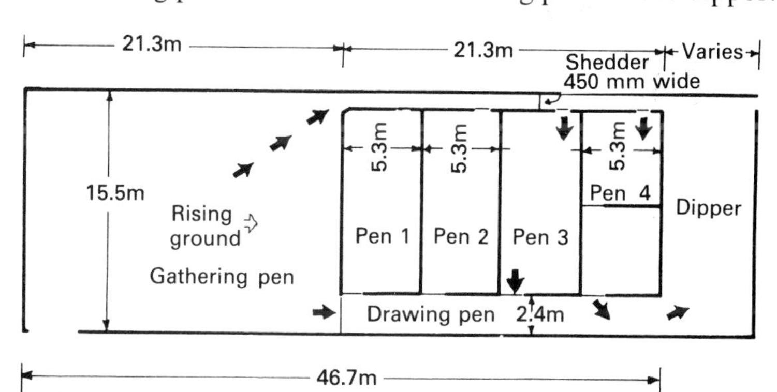

Pens 1, 2, 3, 4 equal in area (approx) to gathering pen
Drawing pen equal in area to one side pen

Figure 9.1. Handling pen layout with side shedder
Source: Shepherd, C.S., 1974

The total area of side pens and drawing pens should equal the area of the gathering pen. All of the sheep gathered initially can then be accommodated after shedding or drawing. The layout is rectangular and would be best sited on ground rising from the entrance of the gathering pen to the dipper draining pens at the top.

Centre shedding race

Sheep are funnelled into the central shedding race from the gathering pen usually through a forcing gate. There are labour saving advantages over the side shedding layout in that one man can shed three ways if the race is at least 5 m long from the entrance to the shedding gate. Spraying, dipping and dosing are done centrally. A further refinement of the centre race layout is the addition of a diamond-shaped forwarding pen between the gathering pen and the shedding race. Sheep flow readily into the parallel handling pens then on to the shedder. This layout has the added advantage that it need not necessarily be rectangular and the shape of the pens can be modified to suit the site.

Plate 41. Prefabricated demountable pens can be moved near to the housing in winter and to the grazing area in summer *Farmers Guardian*

Planning a layout

There are ten basic components in a successful sheep handling unit and each has to be designed in relation to flock size and overall pen layout. The components are:

1. Gathering pen.
2. Side pens.
3. Forwarding pens.
4. Handling race.
5. Shedding race.
6. Foot bath.
7. Drawing pen.
8. Dipper catching pen.
9. Dip bath.
10. Dipper draining pen.

Gathering pen

The capacity of the handling pen should be related to flock size and expected daily throughput for a big operation such as scab dipping: 250 ewes and their lambs or 500 head seems a reasonable day's work to me. Others may be more ambitious, but if each animal is to be immersed for at least 60 seconds, two animals at a time would take four hours of actual dipping. Each ewe and her lambs would need 0.65 square metre of gathering pen or 160 square metres for 250 ewes and their lambs. Most gathering pens will be rectangular, although slope can be varied in relation to obstructions and natural factors, such as roads or walls. The surface of the pen should be free draining as previously specified.

Forwarding pen

Between the gathering pen and the central shedding race a diamond-shaped forwarding pen allows numbers of sheep to be handled before they are passed on to the dosing pen. The shape encourages easy movement of animals into the dosing pen or shedding race; 60 square metres should accommodate 200 ewes at a time.

Double handling race

Two parallel pens of different widths between the forwarding pen and the shedding race facilitate sheep movement in single file into and through the race. It is a very useful facility for close handling small batches of thirty to fifty sheep at one time. If the wide pen holds twenty sheep and the narrow pen ten for dosing, two men can work comfortably along the length of the pen. To ensure smooth flow into the shedding race the exit should be funnel shaped, not less than a 30-degree angle. With a see-through gate in the centre fence one group of sheep can be used to decoy the rest of the flock into

Plate 42. Gates should have positive and easily-operated catches

Farmers Guardian

this pen and on into the race. The whole of this working area should be built on a concrete slab: 100 mm concrete laid over 100–150 mm hardcore with a side fall of 50 mm to aid the surface drainage. The slab should also be 100 mm above the level of the surrounding ground.

Shedding race (Figures 9.1 and 9.4)
Positioning and design of the shedding race is critical to the success of the whole layout. A side shedding race 15–18 m long will allow one man to sort two ways or two men to sort three ways. A centre race on the other hand permits three-way sorting by one man or five-way sorting by two men. Quick and accurate shedding can only be achieved if the operator has an unobstructed view of the entire length of the race, and the sheep must sort themselves into single file at the entrance to the race.

Holding pens
A minimum of four holding pens is needed for quick and efficient sorting from the shedding race. Pen divisions can be

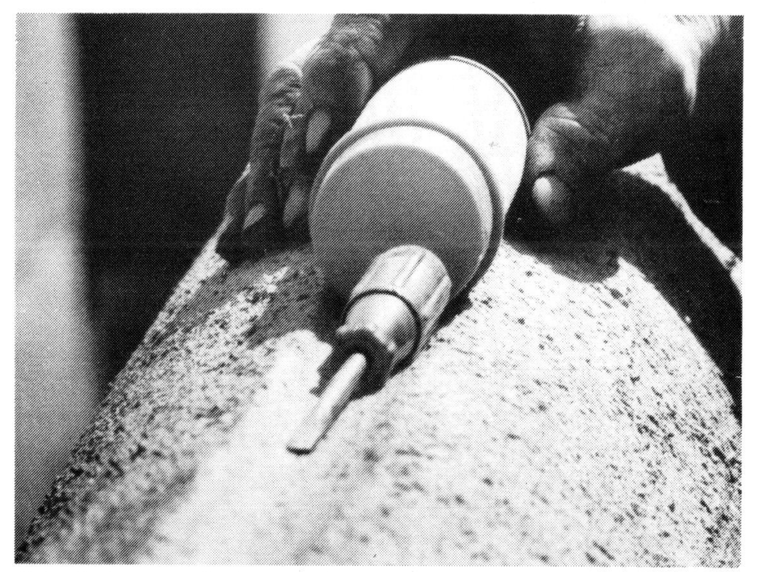

Plate 43. A groove in the top of the handling pen wall used to hold tools, drenches etc *Farmers Guardian*

post and rail fencing and the floor of the pens should be the same specification as the gathering pen.

Drawing pen

The shape and size of drawing pen can vary with the layout, its main function though is to transfer different lots of sheep from the holding pen to the dipper catching pen. Design must allow for the sheep to be drawn directly from the side holding pens and it is usually a side passage 1.8 m–1.4 m wide linking all of the side pens. The drawing pen area should be at least equal that of the largest side pen so that all sheep can be drawn in one operation.

Dip bath catching pen

The dipping bath work area usually consists of a square or circular catching pen, a parallel sided or circular dip bath and two draining pens. Size varies according to the layout and number of sheep to be dipped. The best catching pens are circular and for fifty sheep the inside diameter would be 4.5–5.2 m, the perimeter wall 1.06–1.4 m high, which allows for 0.33 square metres to 0.37 square metres for each sheep. Each of the draining pens should have the same area as the catching pen so that they can hold fifty sheep. Two or three rotating gates hung on a centre post permit easy catching and a good supply of sheep to the dipper; the gates should each have a sliding bolt so that they can be secured at a number of points on the pen wall. Sheep are more easily lowered into the dip bath if a slab of marble is inserted in the concrete at the dip entrance.

Foot bath

If adult ewes are to be thoroughly treated by walking through the foot bath rather than standing in it, a minimum length of 3 m is necessary for the formalin to penetrate to all parts of the foot in the time taken to walk through. For best results the liquid in the bath should be changed frequently. A foot washing bath before the foot bath is an added luxury which would add another 3 m to the length of the race. A timber clad race with a concrete bath is the cheapest and easiest to construct (Figure 9.5). Glassfibre foot baths are available but

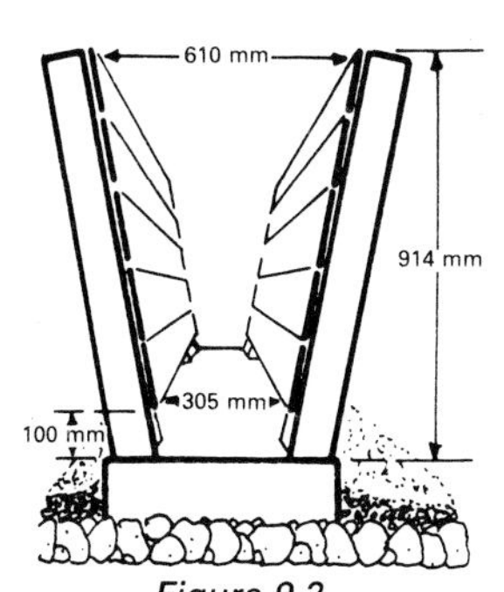

Figure 9.2.
Shedder with tapered sides
Source: Shepherd, C.S., 1974

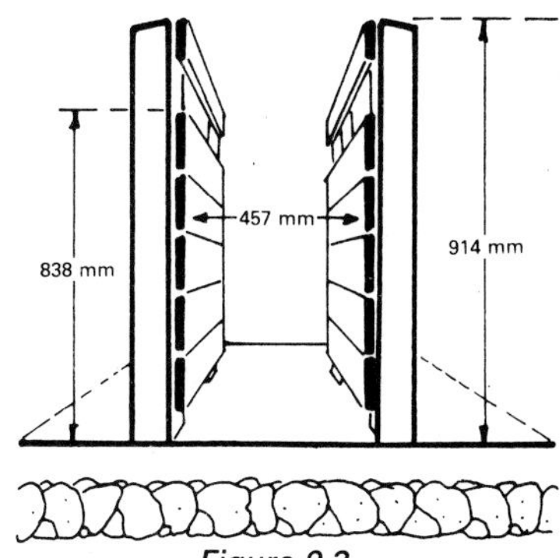

Figure 9.3.
Shedder with parallel sides
Source: Shepherd, C.S., 1974

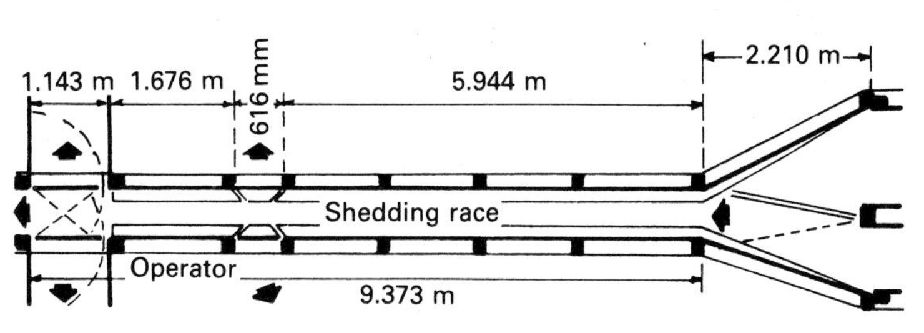

Figure 9.4. Layout of shedding race and gates
Source: Shepherd, C.S., 1974

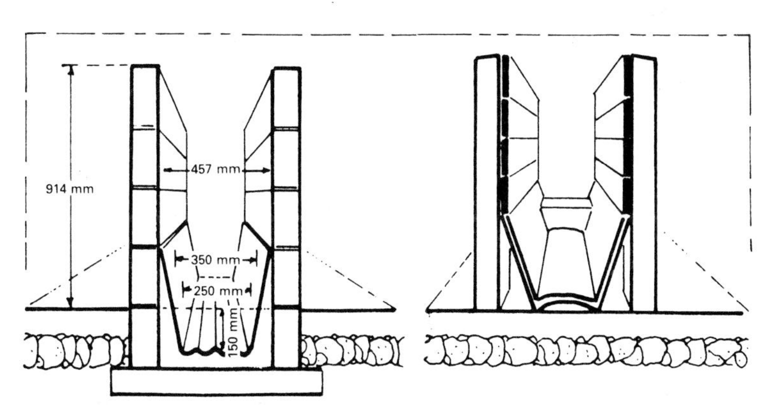

Figure 9.5. Footbath built with concrete blocks
Source: Shepherd, C.S., 1974

Figure 9.6 Foot bath with glassfibre trough
Source: Shepherd, C.S., 1974

generally are not long enough for effective 'walk through' treatment. The concrete floor of the bath should be corrugated to open the claws of the feet. After treatment sheep should exit to a hard standing, usually the dip bath draining pens.

The dipping bath

Total immersion in a plunge bath is the most effective way of dealing with external parasites. Design must allow for the sheep to be plunged beneath the surface of the dip once to wet the head and ears, and for effective scab control sheep must remain in the liquid for at least one minute. Sprays and showers are not as effective as plunge dips and tend to do a much better job of wetting the shepherd rather than the sheep.

Aim for an easy handling and appropriate dip capacity. A good compromise is to allow 2.25 litres a head with a minimum of 364 litres (160 sheep). The volume should be known exactly and marked off in paint on the wall of the bath. Topping-up is necessary after about every 100 sheep or when the level in the bath has dropped by 225 litres. After diping is over remember that the dip is highly toxic to fish and should not be drained into streams, ditches or ponds, therefore it is advisable to incorporate a soakaway or pump the liquid out and spread it in a safe place. If there is the slightest chance of contaminating a water course or spring ask your Regional Water Authority for advice at the planning stage.

Dipping baths may be rectangular, wedge shaped or circular.

Table 9.1. Dip bath dimensions

Type	*Volume (litres)*	*No. of sheep*	*Length (m)*	*Width (mm)*	*Depth (m)*	*Diam. (m)*
Rectangular long swim	1,091	250	3.20	533	1.37	—
Rectangular long swim	1,800	800	5.3	533	1.37	—
Rectangular walk in	—	800	7.62	—	—	—
Wedge shaped short swim	1,091	280	2.74	686-991	1.37	—
Circular	1,800	2,000	—	—	—	1.5

Plate 44. A long swim 'walk in' dip bath *Farmers Guardian*

Perhaps the easiest and cheapest to construct is the parallel sided bath which can be cast in situ from concrete or built from high-density concrete blocks finished with waterproof cement. A rectangular bath to cope with 250 ewes and their lambs will hold 1,100 litres of dip, dimensions are given in Table 9.1. A variation on the rectangular dip bath is the 'walk in dipper'. Sheep are held in a pen with a funnel-shaped exit and enter the bath through a non-return gate. It is claimed that the sheep will walk through the gate, down the ramp and into the dip. They swim to the deepest part where they are pushed under and then they walk up the exit ramp. Some people can make this type of dipper work, I cannot and have always ended up pushing sheep through the non-return gate and then manhandling them into the bath. Check with other users before committing yourself; if it does not work it is a far more tiring system to operate than a circular catching pen. The fibre glass 'walk in' bath is sunk into the ground and backfilled with a weak mix of concrete. Beware, if water seeps between the fibre glass and the concrete, the bath sides may be forced inwards even to the point of collapse.

Wedge-shaped dip baths taper from a wide entrance to a narrow exit, the cross sectional area of the bath also tapers with depth. The extra width at the entrance allows easy turning of sheep and, it has been claimed, better penetration of dip as the wool floats to the surface more readily. More frequent topping up may be needed to maintain the working depth at the surface of the dip.

Circular dipping baths are most suitable where large numbers of sheep are to be dipped; a round bath to handle 2,000

Table 9.2. Guide to economic dipper size

Flock size (ewes)	*Capacity (litres)*	*Type of dipper*
up to 500	900–1,300	short swim
500–1,000	2,000–2,700	long swim or circular
over 1,000	up to 4,500	long swim or circular

Source: WATSON, G.A.L. *et al* (1982).

Plate 45. Circular dip baths are most suitable for large flocks *Farmers Guardian*

sheep would have a volume of 1,800 litres. There should be a sloping exit from the catching pen with a marble slab again, and exit is via a 30-degree sloping ramp. Construction in brick with a waterproof cement lining is complicated and moulded fibre glass baths are available.

Draining pens

After dipping, the sheep should stand in a drainage pen to allow the dip liquid to drain back into the bath through a filter. A one in thirty slope back to the bath is ideal; the floor should be non-slip and two 'side by side' pens are needed.

Fencing and walls

Choice of materials depends very much on their price and availability. Posts should preferably be hardwood in the concrete areas and where softwood is used it should be pressure impregnated with preservative. Post and rail fencing, plywood and metal-clad fences, brick and concrete block walls have all been used. Solid sides are preferable in close handling areas and for the catching pen, otherwise open fencing allows a better circulation of air. Specifications for fencing and cladding are given in Figure 9.7.

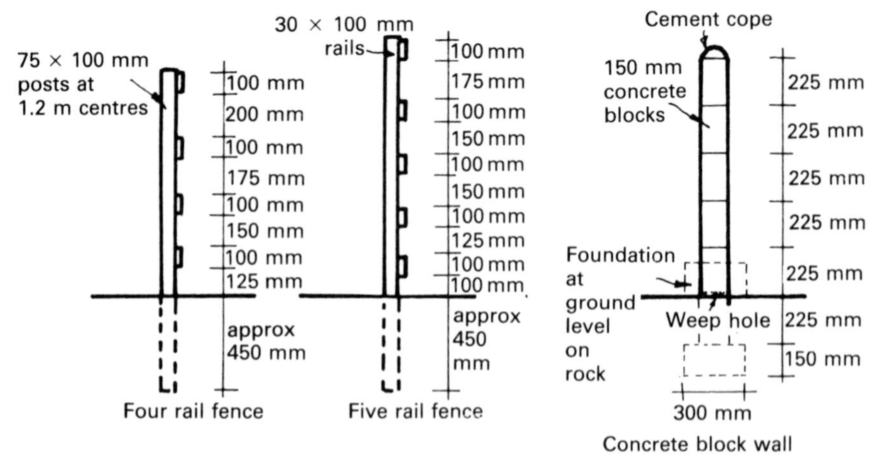

Figure 9.7. Sheep pen fence and wall specifications
Source: Watson, G., et al, 1981

Cost

The layout in Figure 9.8 is suitable for handling 250 ewes. Holding pens are floored with hardcore; catching, dipping, draining, footbath and shedder areas are concrete. Fences are posts and rail, walls are concrete blocks which are cement plastered. The dipping bath and footbath are made from fibre glass. At December 1983, this handling unit cost £9,783 (£22.27 m^2) before grant. The total breakdown by section and trade is given in the 'Farm Building Cost Guide 1984'.

References

MAFF, *Sheep handling pens and baths*, leaflet 14, HMSO, 1960.
SHEPHERD, C.S. (1974), *Design and layout of sheep handling pens,* WOSCA Bull, 159.
WATSON, G.A.L. (1981), *Farm Building Progress 71,* 21-8.

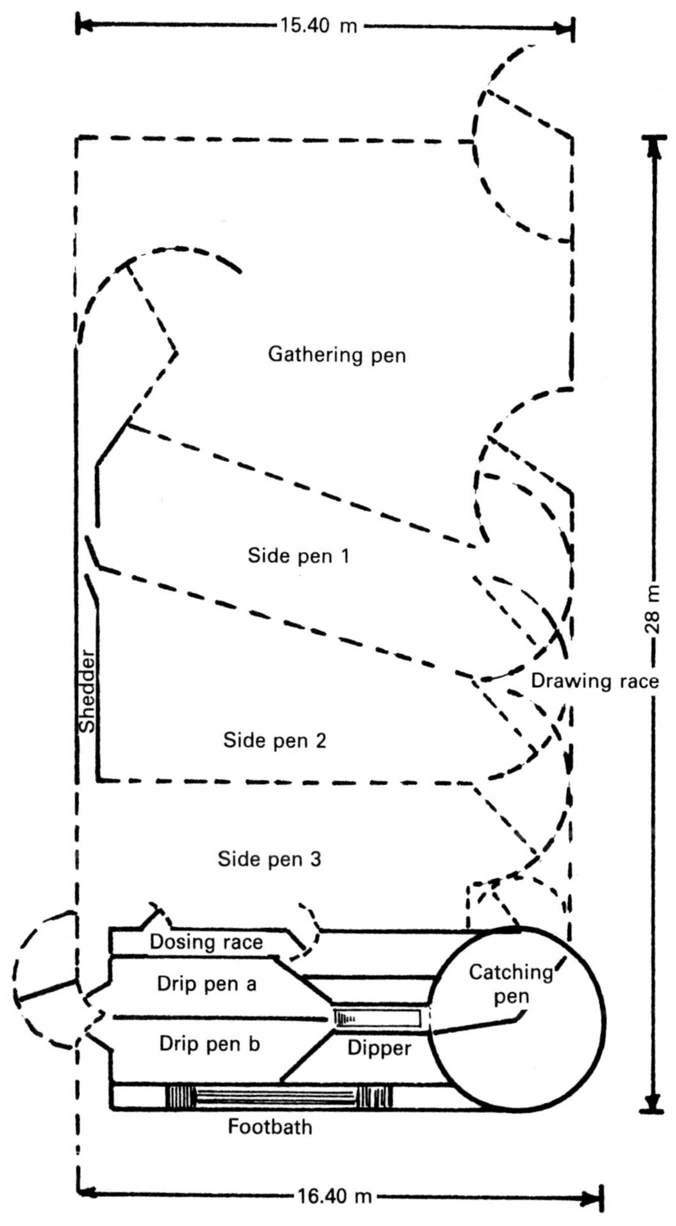

Figure 9.8. Sheep handling pens
Source: SFBIU, 1984, 'Farm Building cost guide'

APPENDICES

Appendix 1

Output per ewe

Output per ewe has been calculated using the following formula.

$$\text{Output} = E\,[a(X - d - e) + gd + c + s].$$

Where E = Number of ewes mated,

a = average lamb sale per ewe mated,

X= lamb: ewe ratio of lambs at weaning per ewe mated,

$$d = \frac{1}{(\text{Lamb crops per ewe})}$$

e = ewe mortality as a percentage of ewes mated,

g = average sale price of cast ewes,

c = average sale of wool per ewe mated,

s = subsidy per ewe mated.

Appendix II

Body condition scoring

'Body condition scoring' has been developed in an attempt to standardise the description of an animal, on a scale from one to five.

Body condition is assessed by pressing with the fingers at positions A and B, the top and sides of the backbone in the loin area behind the last rib.

Follow the following procedure and use it to assess condition on the scale one to five:

1st. Feel the degree of sharpness or roundness of the bony upward pointing spikes (spinous processes) rising from the backbone (Position A).

2nd. Feel the degree of cover over the horizontal bones (transverse processes) coming out from the side of the backbone (Position B).

3rd. Try to pass your fingers under the ends of the transverse processes.

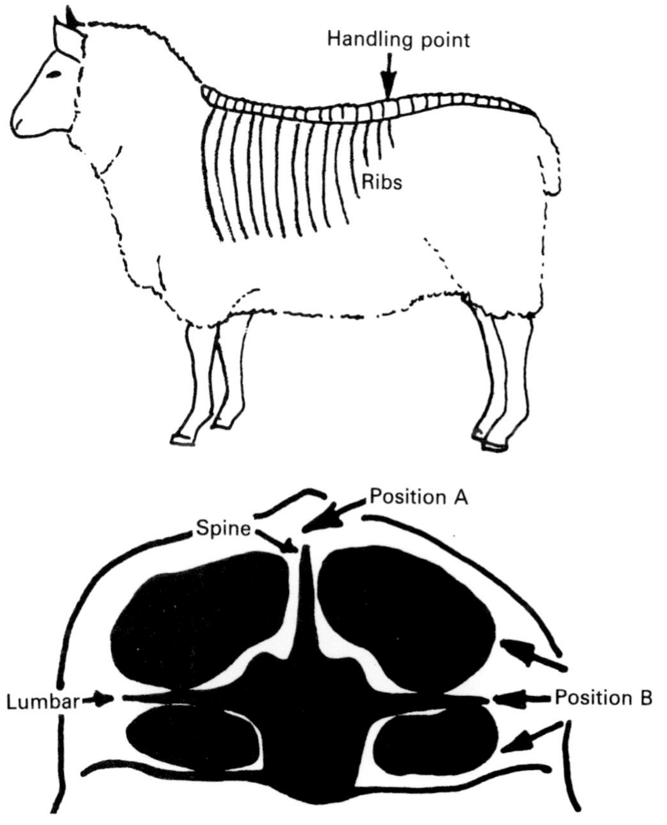

Appendix Figure 1. How to condition score

Source: Hughes, G., 1981

Body Condition Scoring

Score	*Description*	
	Position A *Spinous*	*Position B* *Transverse processes*
1. (emaciated)	Prominent and sharp	Sharp — possible to feel between each bone
2. (lean)	Prominent but smooth	Smooth and rounded
3. (good condition)	Smooth and rounded, pressure needed to feel individual spikes	Smooth and well-rounded, firm pressure needed to feel the ends of the bones
4. (fat)	Considerable presssure needed to feel spines	Ends of bones not detectable
5. (exessively fat)	Spines undetectable, depression due to fat cover	Cannot be felt

INDEX